T0269298

GREEN'S FUNCTIONS AND ORDERED EXPONENTIALS

This book presents a functional approach to the construction, use and approximation of Green's functions and their associated ordered exponentials. After a brief historical introduction, the author discusses new solutions to problems involving particle production in crossed laser fields and non-constant electric fields. Applications to problems in potential theory and quantum field theory are covered, along with approximations for the treatment of color fluctuations in high-energy QCD scattering, and a model for summing classes of eikonal graphs in high-energy scattering problems. The book also presents a variant of the Fradkin representation which suggests a new non-perturbative approximation scheme, and provides a qualitative measure of the error involved in each such approximation. In addition, it deals with adiabatic and stochastic approximations to unitary ordered exponentials.

Covering the basics as well as more advanced applications, this book is suitable for graduate students and researchers in a wide range of fields, including quantum field theory, fluid dynamics and applied mathematics.

H. M. FRIED received his PhD from Stanford University in 1957. He spent a post-doctoral year at the Ecole Normale Supérieure in Paris and then three years teaching physics at UCLA. This was followed by a year as a visiting member of the Institute of Advanced Study in Princeton, and two years as a visiting physicist at the Courant Institute at NYU, before joining the Physics Department at Brown University. Professor Fried has lectured and performed research in university departments and institutes throughout the world, principally in Paris, Marseille and Nice, and is a Director of the Workshops on Non-Perturbative QCD, which alternate between the American University of Paris and La Citadelle, Villefranche-sur-Mer. He is now Professor Emeritus of Physics at Brown University and continues to do occasional teaching there, as well as maintaining a research program in theoretical aspects of quantum field theory.

GREEN'S FUNCTIONS AND ORDERED EXPONENTIALS

H. M. FRIED

CAMBRIDGE
UNIVERSITY PRESS

CAMBRIDGE UNIVERSITY PRESS
Cambridge, New York, Melbourne, Madrid, Cape Town, Singapore, São Paulo

Cambridge University Press
The Edinburgh Building, Cambridge CB2 2RU, UK

Published in the United States of America by Cambridge University Press, New York

www.cambridge.org
Information on this title: www.cambridge.org/9780521443906

First published 2002
This digitally printed first paperback version 2005

A catalogue record for this publication is available from the British Library

Library of Congress Cataloguing in Publication data
Fried, H. M. (Herbert Martin)
Green's functions and ordered exponentials / H. M. Fried.
p. cm.
Includes bibliographical references and index.
ISBN 0 521 44390 3
1. Green's functions. 2. Exponential functions. 3. Mathematical physics. I. Title.
QC174.17.G68 F75 2002
530.15′535 – dc21 2002017417

ISBN-13 978-0-521-44390-6 hardback
ISBN-10 0-521-44390-3 hardback

ISBN-13 978-0-521-44862-8 paperback
ISBN-10 0-521-44862-X paperback

This book is dedicated to the memory of three extraordinary Physicists and Human Beings, men who died during the ten-year period of the writing of this book. From these Scholars and Gentlemen the author was privileged to learn a little of both Physics and the Humanity which can coexist in even the greatest of scientists: Profs. Donald Yennie, Antoine Visconti, and Julian Schwinger.

Contents

Contents

Preface

Physics, and indeed all of Science and rational Life, is a causal affair. Events occur in a well-defined way; and even though nonlinear effects may mask a precise understanding of an underlying mechanism, there can be no rational doubt that cause preceeds effect. The mathematical expression of this truth is couched in the language of Green's functions (GFs), originally invented to provide solutions to electrostatic problems, and subsequently generalized to give compact expression to the causality which appears in time-dependent situations.

At the same time, it has become at least partially clear that when a very large number of iterations of an interaction are associated with the nonlinear, or strong-coupling description of a system, it is not always possible to link specific causes with observed effects. Thus the transition to chaos observed first in the multiple repetition of simple maps, and then in the fractal behavior of physical fluids as they approach fully developed turbulence; thus the realization that strongly coupled gluons and quarks of QCD need not propagate in the causal manner expected from perturbative approximations. Causality is clearly and explicitly true in weakly coupled systems, even though this property can be masked when essential nonlinear dynamics prevent the identification of a specific effect as due to a specific cause.

In recent years, utilization of GF techniques has grown to encompass an immense number of disparate subjects, including application to the large-scale structure of nonlinear systems. Whether one is treating classical or quantum mechanics, Navier–Stokes fluids or ordinary nonlinear differential equations, there is a corps of analogous problems which can advantageously be treated by these methods.

In the general representation and construction of such GFs, encountered across a wide variety of fields, one meets and must deal with ordered exponentials (OEs); and it is for this reason that the latter subject forms an indispensable part of this book. OEs are interesting functions in their own right, but very little is known about their non-perturbative approximations; what is presented here is intuitive, physically motivated, and with a certain connection with low-frequency approximations to nonlinear problems. Other applications, such as the use of OEs to obtain formal solutions to Euler and Navier–Stokes equations, have been left for another occasion.

It must be stated clearly that some of the results stated in this book, for which the author is in part responsible, are without rigorous mathematical foundation. To a physicist, intuition has its own value, which too often becomes its only justification; but it is from this point of view that much of the material of the latter chapters should be understood. At the very least, mathematicians will find in this slim volume a number of intuitively based statements which are in need of rigorous proof, or disproof.

Some of the fundamental topics presented here – such as basic functional methods, and the Schwinger/Fradkin formalism for causal GFs – follow quite closely material appearing in a previous book by the author,[1] called "HMF#2", while some references have been made to material in an even older book[2] by the author, hereinafter called "HMF#1". Including the last Section of Chapter 3, and with the exception of Chapter 7, essentially all of the remaining material presented is new, dating from the past decade.

The level of the present work is again such that graduate students and professionals in mathematical science should find its material and concepts quite familiar. Dirac delta-functions, for example, are used without hesitation; and where all readers may not have a working acquaintance with functional methods, a brief introductory sketch is given, sufficient for the purpose at hand. But the techniques presented are surely applicable to a wide variety of subjects; and each reader, it is hoped, will find a significant measure of success when applying them to his or her own pressing, nonlinear problems.

This book was begun during the academic year 1991–92, when the author was a Visiting Professor at the Université de Nice; and completed slowly over the following nine years at Brown University. To friends and colleagues of both institutions are due the warmest thanks and acknowledgement of many kindnesses. Comme avant, je leur remercie de tout.

Brown University   *H. M. Fried*

Notes

1 *Functional Methods and Eikonal Models*, Éditions Frontières, Gif-sur-Yvette, France (1990), hereinafter referred to as HMF#2
2 *Functional Methods and Models in Quantum Field Theory*, The MIT Press, Cambridge, MA (1972), hereinafter referred to as HMF#1

Abbreviations

CM	Center of Mass
DE	Differential Equation
FFT	Functional Fourier Transform
FI	Functional Integral
$\hat{G}F$	Generating Functional
GF	Green's Function
IR	Infrared
LHP	Lower Half Plane
LHS	Left-Hand-Side
MSA	Mass-Shell Amputation
ODE	Ordinary Differential Equation
OE	Ordered Exponential
QCD	Quantum Chromodynamics
QED	Quantum Electrodynamics
QFT	Quantum Field Theory
RHS	Right-Hand-Side
SC	Strong Coupling
SWE	Schrödinger Wave Equation
UHP	Upper Half Plane
UOE	Unitary Ordered Exponential

1

Introduction

1.1 Historical remarks

It is difficult to fix the precise beginning of the vast and disparate subject matter which now exists under the name of "Green's Functions", but the origins of the method may certainly be associated with the original and ingenious work of George Green (1793–1841).[1] That application, now called Green's Theorem, of Gauss' Theorem applied to electrostatics, in modern language makes use of the differential statement

$$\nabla^2 |\mathbf{r} - \mathbf{r}'|^{-1} = -4\pi \delta(\mathbf{r} - \mathbf{r}'). \tag{1.1}$$

Before the advent of the Dirac delta-function, the content of (1.1) had to be expressed in a somewhat circuitous way,[2] which is how Green treated the problem.

Every modern text on potential theory begins with the statement of Gauss' Theorem, $\int d\mathbf{S} \cdot \mathbf{F} = \int d^3\mathbf{r}' \nabla' \cdot \mathbf{F}(\mathbf{r}')$, where $\mathbf{F}(\mathbf{r}')$ is a continuous and differentiable vector function, whose divergence is to be integrated over a volume bounded by the surface $\int d\mathbf{S}$. Green noted that the choice $\mathbf{F} = V\nabla U - U\nabla V$ generates, for arbitrary U, V,

$$\int d\mathbf{S}' \cdot [V\nabla' U - U\nabla' V] = -\int d^3r'[U(\mathbf{r}')\nabla'^2 V(\mathbf{r}') - V(\mathbf{r}')\nabla'^2 U(\mathbf{r}')],$$

$$\tag{1.2}$$

which, in vector notation, is the statement of Green's theorem. If the further choice $U(\mathbf{r}') = -(4\pi |\mathbf{r} - \mathbf{r}'|)^{-1}$ is made, where $\mathbf{r}$ denotes the radius vector (drawn from an origin of arbitrary location) of a point inside the integration volume, then (1.1) and (1.2) yield

$$V(\mathbf{r}) = -\frac{1}{4\pi} \int d^3r' \frac{1}{|\mathbf{r} - \mathbf{r}'|} \nabla'^2 V(\mathbf{r}')$$

$$+ \frac{1}{4\pi} \int d\mathbf{S}' \cdot \left[\frac{1}{|\mathbf{r} - \mathbf{r}'|} \nabla' V(\mathbf{r}') - V(\mathbf{r}')\nabla' \frac{1}{|\mathbf{r} - \mathbf{r}'|} \right]. \tag{1.3}$$

1

If $V(\mathbf{r})$ now refers to the electrostatic potential due to a specified charge distribution at points within the surface,

$$\nabla^2 V(\mathbf{r}) = -4\pi\rho(\mathbf{r}), \tag{1.4}$$

then (1.3) provides an expression for $V(\mathbf{r})$ given in terms of quadratures over the "Green's function" $G(\mathbf{r} - \mathbf{r}') = U(\mathbf{r} - \mathbf{r}')$ multiplied by the charge density, to which must be added the contributions of the surface integrals of (1.3) over values of V and/or ∇V that are specified as boundary conditions. In other words, the solution to (1.4) may be written as

$$V(\mathbf{r}) = -4\pi \int d^3r' G(\mathbf{r} - \mathbf{r}')\rho(\mathbf{r}'), \tag{1.5}$$

to which must be added the RHS surface terms of (1.3). As long as $\mathbf{r}$ does not lie on $\int d\mathbf{S}$, these surface terms satisfy the homogeneous equation of Laplace, while the volume integral of (1.5) generates a solution to the inhomogenous equation (1.4) of Poisson.

This structure, of (1.5) plus appropriate solutions of the homogeneous equation, has over the intervening two centuries been generalized from the relatively straightforward elliptic (1.4) to hyperbolic and partial differential equations (DEs), and to nonlinear problems such as those of Navier–Stokes fluids and quantum field theory (QFT). In each case, the solution of an inhomogeneous DE in n spatial dimensions,

$$\mathcal{D}\phi(\mathbf{r}, t) = j(\mathbf{r}, t), \qquad \mathcal{D} = \mathcal{D}\left[\frac{\partial}{\partial t}, \nabla; A(\mathbf{r}, t)\right], \tag{1.6}$$

specified by some collection of differential operators and (in the nonlinear case) associated fields $A(\mathbf{r}, t)$, is given by

$$\phi(\mathbf{r}, t) = \int d^n r' \int_{-\infty}^{+\infty} dt' G(\mathbf{r}, \mathbf{r}'; t, t'|A)j(\mathbf{r}', t') + S(\mathbf{r}, t), \tag{1.7}$$

where the $S(\mathbf{r}, t)$ specify needed boundary and/or initial conditions of the problem, and are solutions of the homogeneous relation $\mathcal{D}S = 0$. The Green's function (GF) of the problem, $G(\mathbf{r}, \mathbf{r}'; t, t'|A) = \langle \mathbf{r}, t|\mathcal{D}^{-1}|\mathbf{r}', t'\rangle$, is a solution of the relevant generalization of the inhomogeneous (1.1),

$$\mathcal{D}G = \delta(\mathbf{r} - \mathbf{r}')\delta(t - t'). \tag{1.8}$$

In this way, Green's original formulation of general solutions to electrostatic problems has found a natural generalization to virtually all fields whose essential Physics is described by an inhomogeneous DE.

In subsequent sections, specific forms for $\mathcal{D}^{-1}$ will be given for problems of interest in fluid motion and diffusion, whose underlying symmetry is Galilean;

and for the propagators of QFT, of Lorentzian symmetry. Attention will be focused mainly on hyperbolic DEs, requiring time-dependent initial conditions; and simple constructions illustrating the method of enforcing different initial conditions will be described. These relatively simple computations are associated with solutions of a linear problem, and such techniques can provide only formal descriptions of nonlinear, or interacting systems, where $\mathcal{D}$ is a function of fields A that are to be coupled (by means of other equations) to the desired solution ϕ.

A more explicit construction of $\mathcal{D}^{-1}$ in the presence of external interactions will also be given in terms of the exact, and most useful representation of Fradkin.[3] Special variants of the Fradkin representation generate a new, non-perturbative method for exact and approximate representations of these GFs; and in these approximations, one has at least a qualitative idea of their error. For vectorial interactions, one learns in Chapter 6 of possible chaos appearing in the realization of such non-perturbative approximations; and one sees just how such chaos is naturally removed in QFT, which process suggests application to methods of chaos suppression for classical systems. One learns, in the context of any Fradkin representation, the intimate connection between such GFs and ordered exponentials (OEs), which leads, in Chapter 9, into a discussion of unitary OEs. A brief discussion of known methods of extracting the infrared, or low-frequency structure of relevant GFs is given in Chapter 7, while a solution for the "scalar laser" problem of Chapter 4 is used in Chapter 8 to construct a model GF which can be used to estimate the total cross section for particle production in a "modified multiperipheral model" at extreme, relativistic energies. A new solution for pair production in the presence of a non-constant electric field is described in Chapter 3, while estimates are given in Chapter 4 for the same process in the overlapping fields of two high-intensity lasers. Some of these results are old, and some are new; but all can be given a succinct description in terms of GFs and OEs.

1.2 Linear Physics

In this section will be described the simplest linear prototypes of propagator found in four distinct fields: non-relativistic fluid motion, the non-relativistic Schrödinger equation, ordinary DEs, and QFT. Motion associated with a simple harmonic oscillator driven by an arbitrary source $g(t)$ is the simplest ODE imaginable,

$$\frac{\mathrm{d}^2 x}{\mathrm{d}t^2} + \omega^2 x = g(t), \tag{1.9}$$

to be solved, for definiteness, under the initial conditions $x(0) = D$, $\mathrm{d}x(0)/\mathrm{d}t = 0$. More complicated problems of current interest are obtained by inserting damping and, for example, replacing ω^2 by $\omega^2(x^2 - 1)$ to produce the Duffing equation, with its manifest nonlinear behavior. What shall be done in this section is to generate solutions to the linear problems using standard GF methods, and then to compare the results with an alternative and equivalent phase-space method of solution. No OEs appear in the linear analysis, but the standard questions of retarded or advanced, causal or anticausal solutions must be answered.

Adding spatial derivatives to (1.9) in a Lorentz-symmetric way generates the forms of non-interacting field theory,

$$(\mu^2 - \partial^2)A(x) = j(x), \tag{1.10}$$

where causality will be demanded in the sense that $A(\mathbf{r}, t)$ cannot be different from zero until a signal from the source $j(\mathbf{r}, t)$ (traveling at the speed of light when $\mu = 0$) can reach the point $x = (r, t)$; here, ∂^2 denotes the d'Alembertian operator, $\nabla^2 - \frac{1}{c^2}\frac{\partial^2}{\partial t^2}$, and units will be chosen in what follows such that c, the velocity of light, is unity.

In contrast, the diffusion equation of (relatively) low-velocity fluid motion is non-relativistic,

$$\left(\frac{\partial}{\partial t} - \nu\nabla^2\right)\mathbf{v} = \mathbf{f}(\mathbf{r}, t), \tag{1.11}$$

where ν denotes viscosity and $\mathbf{v}(\mathbf{r}, t)$ is the fluid velocity; appropriate initial conditions here will again demand causality. Because it has but one time derivative, there exist but two GF solutions for this problem, one "retarded" (subscript R) and the other "advanced" (subscript A); and it is simplest to begin the detailed construction of these GF s with this example.

(i) Non-relativistic diffusion: The requirement of causality will select the GF G_R as the physically relevant solution of the inhomogeneous

$$\left(\frac{\partial}{\partial t} - \nu\nabla^2\right)G_R(\mathbf{r} - \mathbf{r}'; t - t') = \delta(\mathbf{r} - \mathbf{r}')\delta(t - t'), \tag{1.12}$$

which, as written in (1.12), turns out to be a function of coordinate differences.

If one knows the general solutions to the corresponding homogeneous DE, the solution to (1.12) may be expressed as a summation over all eigenstates of positive eigenvalues E_n, in the form

$$G_R(\mathbf{r} - \mathbf{r}'; t - t') = \theta(t - t')\sum_n u_n(\mathbf{r})u_n^*(\mathbf{r}')\exp[-E_n(t - t')], \tag{1.13}$$

where the $u_n(\mathbf{r})$ form a complete orthonormal set, satisfying $[E_n + \nu\nabla^2]u_n = 0$ and $\sum_n u_n(\mathbf{r})u_n^*(\mathbf{r}') = \delta(\mathbf{r} - \mathbf{r}')$. The θ-function of (1.13) expresses the retardedness of the GF; because $\theta(x) = 1$, $x > 0$, and $\theta(x) = 0$, $x < 0$, the solution to (1.11),

$$\mathbf{v}(\mathbf{r}, t) = \int d^3 r' \int dt' G_R(\mathbf{r} - \mathbf{r}'; t - t') \mathbf{f}(\mathbf{r}', t') + \mathbf{v}_0(\mathbf{r})$$

will have no contribution to its first RHS term for $t' > t$, so that an effect at t cannot appear before its generation at t'. Here, $\mathbf{v}_0(\mathbf{r})$ represents the initial condition of this problem, the velocity field specified at all points $\mathbf{r}$, and satisfying the equation $\nabla^2 \mathbf{v}_0(\mathbf{r}) = 0$. Mathematically speaking, $\theta(x)$ is really the limit of a sequence of functions chosen such that $\theta(0) = 1/2$; everywhere in this book it may be represented by the integral

$$\theta(x) = \frac{1}{2\pi i} \int_{-\infty}^{+\infty} d\omega (\omega - i\epsilon)^{-1} e^{i\omega x}, \tag{1.14}$$

and its properties checked by straightforward contour integration, as well as by the more conventional relation: $\delta(x) = d\theta/dx$.

It should be noted that the E_n are positive, and hence the summation of (1.13) is sensible; here, the $u_n(\mathbf{r})$ are just plane-wave exponentials of wave-vector $\mathbf{k}_n = 2\pi \mathbf{n}/L$, where $\mathbf{n}$ is a vector each of whose components are integers, L^3 is an appropriate normalization volume, and $E_n = vk_n^2$. Were the viscosity continued to an imaginary value, with a change of normalization, one would be dealing with the non-relativistic Schrödinger wave equation (SWE), as in (ii) below.

The simplest method of construction for any such GF of a linear problem is to employ a Fourier representation,

$$G_R(\mathbf{r} - \mathbf{r}'; t - t') = (2\pi)^{-4} \int d^3 k \int d\omega \tilde{G}_R(\mathbf{k}, \omega) e^{i\mathbf{k}\cdot(\mathbf{r}-\mathbf{r}') - i\omega(t-t')}, \tag{1.15}$$

where $\tilde{G}_R$ is determined by substituting (1.15) into (1.12), and is clearly given by $\tilde{G}_R = i(\omega + ivk^2)^{-1}$. In the complex ω-plane, the integrand of (1.15) has but one singularity, a pole at $\omega = -ivk^2$, which multiplies the factor $\exp[-i\omega(t - t')]$. Evaluating the ω-integral by contour integration, one is forced for $t - t' < 0$ to close the contour in the upper half ω-plane, which yields zero, while the choice $t - t' > 0$ requires closing the contour in the lower half plane, which yields

$$G_R(\mathbf{r} - \mathbf{r}'; t - t') = \theta(t - t')(2\pi)^{-3} \int d^3 k \exp[i\mathbf{k} \cdot (\mathbf{r} - \mathbf{r}') - vk^2(t - t')]. \tag{1.16}$$

Because one finds a non-zero value only for $t - t' > 0$, as expressed by the θ-function of (1.16), this is the retarded GF; G_A would have been obtained by reversing the sign of the $-i\omega(t - t')$ phase of (1.15). Note that in the limit of zero viscosity, (1.16) reduces to $\theta(t - t')\delta(\mathbf{r} - \mathbf{r}')$, so that the solution to (1.13) is simply

$$\mathbf{v}(\mathbf{r}, t) = \int_0^t dt' \mathbf{f}(\mathbf{r}, t') + \mathbf{v}_0(\mathbf{r}).$$

As a Gaussian integral, (1.16) may be evaluated immediately,

$$G_R(\mathbf{r} - \mathbf{r}'; t - t') = \theta(t - t')[4\pi \nu(t - t')]^{-3/2} \exp[-(\mathbf{r} - \mathbf{r}')^2/4\nu(t - t')].$$
$$(1.17)$$

For small differences $t - t'$, most of the contribution to the $\mathbf{r}'$ integral over (1.17) comes from the points $\mathbf{r}'$ close to $\mathbf{r}$; but as the time difference increases, more and more of the source dependence at other values of $\mathbf{r}'$ is available to influence the behavior of the solution at $\mathbf{r}$.

If the source dependence is chosen to be that of a delta-function in space and time, $\mathbf{f}(\mathbf{r}, t) = \mathbf{f}_0 \delta(\mathbf{r}) \delta(t - \epsilon)|_{\epsilon \to 0}$, and if the initial field $\mathbf{v}_0 = 0$, the integrations over (1.17) yield the simple result

$$\mathbf{v}(\mathbf{r}, t) = \theta(t) \mathbf{f}_0 (4\pi \nu t)^{-3/2} \exp[-\mathbf{r}^2/4\nu t]. \qquad (1.18)$$

This has the form of an initial (at $t = 0$) and localized (at $\mathbf{r} = \mathbf{0}$) disturbance diffusing away to all spatial points at later times, while maintaining a constant infinite-spatial integral, $\int d^3 r \mathbf{v}(\mathbf{r}, t) = \theta(t) \mathbf{f}_0$.

It is worth emphasizing the physically intuitive role of such a localized source in "generating" a solution to the homogeneous equation $(\frac{\partial}{\partial t} - \nu \nabla^2)\mathbf{v} = 0$. One imagines for $t < 0$ a perfectly quiescent situation with $\mathbf{v}(\mathbf{r}, t) = \mathbf{0}$, suddenly disturbed by a source $\mathbf{f}_0 \cdot \delta(\mathbf{r}) \delta(t)$ rapidly "turning on and off" at $t = 0$, and generating for subsequent, positive t, a solution to the homogeneous equation. This use of a highly-localized source has long appeared in QFT, as a way of representing the appropriate wave functions of particles "produced" at a particular space–time point. It is an artifice, but a convenient one, and has been used in certain fluid/vortex problems.[4]

(ii) Non-relativistic Schrödinger wave equation: The GFs of interest here are typically "ingoing" or "outgoing", with the latter chosen to represent the "scattered wave" of probability amplitude needed to describe the physical scattering of two particles which interact with each other by means of a potential $V(\mathbf{r}_1 - \mathbf{r}_2)$. Equivalently, by a simple transformation to the CM coordinates of these particles, one may treat the scattering of a particle of reduced mass $\mu = m_1 m_2/(m_1 + m_2)$ in the field of a fictitious potential field $V(\mathbf{r})$, where $\mathbf{r} = \mathbf{r}_1 - \mathbf{r}_2$. In such scattering problems, the energy E of the system is

conserved and specified in advance, so that the original SWE

$$i\hbar\frac{\partial}{\partial t}\psi = \left\{-\frac{\hbar^2}{2\mu}\nabla^2 + V(\mathbf{r})\right\}\psi \qquad (1.19)$$

is replaced by

$$E u(\mathbf{r}) = \left\{-\frac{\hbar^2}{2\mu}\nabla^2 + V(\mathbf{r})\right\}u(\mathbf{r}), \qquad (1.20)$$

upon using the substitution $\psi(\mathbf{r}, t) = u(\mathbf{r})\exp[-iEt/\hbar]$; the corresponding GFs of this problem are to satisfy

$$\left\{E + \frac{\hbar^2}{2\mu}\nabla^2 - V(\mathbf{r})\right\}G(\mathbf{r}, \mathbf{r}'|V) = \delta(\mathbf{r} - \mathbf{r}'), \qquad (1.21)$$

and scattering amplitudes $u(\mathbf{r})$ are given by

$$u(\mathbf{r}) = u_0(\mathbf{r}) + \int d^3r' G(\mathbf{r}, \mathbf{r}'|V)V(\mathbf{r}')u_0(\mathbf{r}'), \qquad (1.22)$$

where $u_0(\mathbf{r})$ represents the "asymptotic input" function $\exp[i\mathbf{k}\cdot\mathbf{r}]$, corresponding to an incident particle of energy $E = \hbar^2k^2/2\mu$ moving in the $\mathbf{k}$ direction.

A formal solution for the GF of this problem may be written as

$$G(\mathbf{r}, \mathbf{r}'|V) = \sum_n u_n(\mathbf{r})u_n^*(\mathbf{r}')(E - E_n)^{-1}, \qquad (1.23)$$

where the summation is over all states of the complete, orthonormal set $u_n(\mathbf{r})$ which satisfy the time-independent SWE,

$$E_n u_n = \left\{-\frac{\hbar^2}{2\mu}\nabla^2 + V(\mathbf{r})\right\}u_n(\mathbf{r}).$$

The RHS of (1.21) is reproduced, upon substitution of (1.23) into (1.21), because the u_n are again assumed complete, satisfying

$$\sum_n u_n(\mathbf{r})u_n^*(\mathbf{r}') = \delta(\mathbf{r} - \mathbf{r}').$$

Typically, the states $|n\rangle$ form a continuum of positive-energy scattering states, plus discreet bound states of negative energy. Note from (1.23) that if the scattering energy E is continued to negative values, one may expect to find a pole of the scattering amplitude when E approaches one of the isolated, negative, bound-state energies. Note also that, because of the spatial dependence of $V(\mathbf{r})$, this GF is no longer a function of the difference of its coordinates. It will become clear below that the distinction between the solutions G_{out} and G_{in} can be made explicit by appending an infinitesimal $\pm i\epsilon$ to the denominator E of (1.23).

8 *1 Introduction*

The exact GFs of this problem cannot be obtained in closed form for an arbitrary $V(r)$, but they can be given an explicit Fradkin representation, from which physically interesting approximations may be drawn. In this "linear" section, however, what is of interest are the GFs defined in the absence of any potential interaction, $G(\mathbf{r}, \mathbf{r}'|0) \to G(\mathbf{r} - \mathbf{r}')$, and again the simplest route is via a Fourier representation,

$$G(\mathbf{r} - \mathbf{r}') = (2\pi)^{-3} \int d^3k \tilde{G}(\mathbf{k}) e^{i\mathbf{k}\cdot(\mathbf{r}-\mathbf{r}')}, \qquad (1.24)$$

and substitution into the $V = 0$ version of (1.21) determines $\tilde{G}(\mathbf{k})$ to be given by $(2\mu/\hbar^2)(k_0^2 - k^2)^{-1}$, where the incident wave-number $k_0 = (2\mu E/\hbar^2)^{1/2}$.

However, because G is singular when $k = (k^2)^{1/2}$ equals $\pm k_0$, one must provide a prescription of how these poles which lie on the path of the k integration are to be avoided. Equivalently, and much more simply, one can displace these singularities by an infinitesimal amount into the complex k-plane, perform the integration along the real k axis, and then set equal to zero the small parameter ϵ used to displace the poles. The two possibilities here correspond to adding $\pm i\epsilon$ to the k_0^2 of $\tilde{G}^{-1}$, and then taking the limit $\epsilon \to 0+$ after the integration; and they will define the two distinct GFs, G_{in} and G_{out}.

There are now two equivalent methods of evaluation, of which one is to be greatly preferred for relativistic problems; in order to make this point we begin with the less preferable but far more common approach. One writes $\tilde{G}^{-1}$ as $-(\frac{\hbar^2}{2\mu})[k^2 - (k_0 \pm i\epsilon)^2]$, since $2i\epsilon k_0$ is equivalent to $i\epsilon$. For definiteness, let us choose the upper sign, so that the poles appear when $k^2 = (k_0 + i\epsilon)^2$, or $k = \pm(k_0 + i\epsilon)$. Integration over solid angle of the factor $\exp[i\mathbf{k}\cdot\mathbf{R}]$, with $\mathbf{R} = \mathbf{r} - \mathbf{r}'$, yields $4\pi \sin(kR)/kR$, which provides an integrand symmetric in k and $-k$. The integral $\int dk$ from 0 to $+\infty$ can then be rewritten as one-half of the same integrand from $-\infty$ to $+\infty$, and the factor of $\sin(kR)$ written as $[\exp(ikR) - \exp(-ikR)]/2i$, with contour integration over the first and second of these terms necessarily closed in the upper and lower k-planes, respectively. Both contributions are the same, and yield the result

$$G_{\text{out}}(\mathbf{r} - \mathbf{r}') = -(\mu/2\hbar^2)\frac{e^{ik_0|\mathbf{r}-\mathbf{r}'|}}{|\mathbf{r} - \mathbf{r}'|}, \qquad (1.25)$$

which for large $\mathbf{r}$ displays a phase growing as $+ik_0\mathbf{r}$, corresponding to an outwardly-moving wave front. The other possibility, G_{in} with a phase of the form $-ik_0|\mathbf{r} - \mathbf{r}'|$, is obtained by adding to the k_0^2 of G^{-1} the infinitesimal $-i\epsilon$.

The alternative method works directly with the Fourier representation of (1.24),

$$G_{\text{out}}(\mathbf{r} - \mathbf{r}') = (2\pi)^{-3}\left(\frac{-2\mu}{\hbar^2}\right) \int d^3k \left(k^2 - k_0^2 - i\epsilon\right)^{-1} e^{i\mathbf{k}\cdot(\mathbf{r}-\mathbf{r}')}. \quad (1.26)$$

One writes an exponential representation for the denominator factor of (1.26), of the form

$$(A - i\epsilon)^{-1} = i \int_0^\infty ds\, e^{-is(A-i\epsilon)} \to i \int_0^\infty ds\, e^{-isA}, \qquad (1.27)$$

where the quantity A is real. Note that, in order to have a sensible integrand for large values of s, the sign in the exponential is determined by the sign of $i\epsilon$ in the denominator of (1.27); for G_{in}, one must use the representation which is the complex conjugate of (1.27). But now the k-integral is Gaussian and immediately performed,

$$\int d^3k \exp[-isk^2 + i\mathbf{k} \cdot \mathbf{R}] = \left(\frac{\pi}{is}\right)^{3/2} \exp[iR^2/4s].$$

The s-integration of (1.27) yields the appropriate Bessel/Hankel function representation,

$$i\pi (2k_0/R)^{1/2} H_{1/2}^{(1)}(Rk_0) \exp[i\pi/4],$$

which when combined with the remaining factors reproduces (1.25). The utility of this method becomes apparent in relativistic problems, when exponentiation is performed using a "proper time" variable s, for the manifest Lorentz invariance of the GF is maintained at every step.

(iii) Ordinary differential equations: We now return to (1.9) and the initial conditions stated there, and solve that problem by a GF method, writing

$$x(t) = \int_{-\infty}^{+\infty} dt'\, G(t - t')g(t') + S(t), \qquad (1.28)$$

where $S(t)$ is a solution of the homogeneous ($g = 0$) form of (1.9). Again, the GF is to satisfy an inhomogeneous "source" equation, analogous to (1.21),

$$\left[\frac{d^2}{dt^2} + \omega^2\right] G(t - t') = \delta(t - t'), \qquad (1.29)$$

and is given a Fourier representation

$$G(t - t') = (2\pi)^{-1} \int dk_0 \tilde{G}(k_0) e^{-ik_0(t-t')}, \qquad (1.30)$$

which, when substituted into (1.29) yields the form of $\tilde{G}$,

$$\tilde{G}(k_0) = -\left[k_0^2 - \omega^2\right]^{-1}. \qquad (1.31)$$

Again, (1.31) is incomplete until one specifies just how its singularities at $k_0 = \pm\omega$ are to be treated, and each of the possible choices defines a distinct solution of (1.29). In this "quadratic" problem there are four independent functions which can be constructed by moving the poles infinitesimally into the complex k_0-plane: both "up", both "down", one "up" and the other "down", and the converse. We shall use the relativistic notation familiar from QFT to label these possibilities as G_A, G_R, G_c, and G_{ac}, respectively; and will make use of the $\pm i\epsilon$ prescriptions to introduce these contour shifts in the simplest way.

These prescriptions can most easily be defined by adding to $\tilde{G}^{-1}$ either the infinitesimals $\pm i\epsilon$, or the terms $\pm i\epsilon s(k_0)$, where $s(k_0)$ denotes the sign of k_0, $s(k_0) = k_0/|k_0|$. For example, if $-\tilde{G}^{-1}$ is replaced by $[k_0^2 - \omega^2 + i\epsilon s(k_0)]$, or (since $\pm 2\epsilon |k_0|$ is equivalent as $\epsilon \to 0$ to $\pm\epsilon$) by $[(k_0 + i\epsilon)^2 - \omega^2]$, the complex poles of this denominator occur when $k_0 = \pm\omega - i\epsilon$. Since contour integration of $\int dk_0$ must be closed in the UHP when $t' > t$, and since these poles lie in the LHP, one can have a non-zero result only for $t > t'$, when the contour must be closed in the LHP. One immediately finds

$$G_R(t - t') = \theta(t - t') \cdot \omega^{-1} \cdot \sin(\omega[t - t']), \tag{1.32}$$

where the subscript R, denoting the retarded GF, has been appended to this distinct solution of (1.29). If the choice $[k_0^2 - \omega^2 - i\epsilon s(k_0)]$ is used for $-\tilde{G}^{-1}$, both poles are displaced into the UHP, leading to a non-zero result only when $t' > t$, which is easily computed to be the advanced GF,

$$G_A(t - t') = \theta(t' - t) \cdot \omega^{-1} \cdot \sin(\omega[t' - t]). \tag{1.33}$$

The Fourier representations for these two independent possibilities differ from each other by the change of sign of k_0, which difference can be removed by the interchange of t and t', leading to (1.33).

Consider now the replacement of $-\tilde{G}^{-1}$ by $[k_0^2 - \omega^2 + i\epsilon]$, which is equivalent to $[k_0^2 - (\omega - i\epsilon)^2]$ and displays poles when $k_0 = \pm(\omega - i\epsilon)$. Contour integration then extracts contributions from both regions $t > t'$ and $t' > t$, and one finds that

$$G_c(t - t') = \left(\frac{i}{2\omega}\right) \exp[-i\omega|t - t'|] \tag{1.34}$$

represents the "causal" GF of this problem, which displays for any non-zero values of $t - t'$ a "positive-frequency time dependence" of a particle "moving forward in time", or (in relativistic field theory) that of an "antiparticle moving backwards in time". Finally, replacement of $-\tilde{G}^{-1}$ by $[k_0^2 - \omega^2 - i\epsilon]$ leads to the "anticausal" GF,

$$G_{ac}(t - t') = -\left(\frac{i}{2\omega}\right) \exp[i\omega|t - t'|], \tag{1.35}$$

which is just the complex conjugate of (1.34).

Which of these GFs should be used for the solution of (1.28)? The answer is that it makes no difference, because the difference of any two of these four GF-solutions to the inhomogeneous (1.29) is a solution of the corresponding homogeneous equation, and as such will appear as part of the term denoted in (1.28) by $S(t)$. In QFT one uses either the retarded, or (more usually) the causal GF, because the requirement of unitarity in a typical scattering problem is then made simpler; in the present case, simplicity suggests using the retarded GF of (1.32), so that

$$x(t) = \omega^{-1} \int_{-\infty}^{t} dt' \sin[\omega(t - t')] \cdot g(t') + S(t), \qquad (1.36)$$

where the solution of the homogeneous problem must have the form: $S(t) = A\cos(\omega t) + B\sin(\omega t)$, and the constants A, B are determined by fitting the initial conditions $x(0) = D$, $dx(0)/dt = 0$. One then, finally, obtains the desired solution

$$x(t) = D\cos(\omega t) + \omega^{-1} \cdot \int_{0}^{t} dt' g(t') \sin[\omega(t - t')]. \qquad (1.37)$$

It is worth commenting on the two independent solutions which exist in this problem to the homogeneous equation, for obvious generalizations can be useful in cases where the complexity of the original equation is considerably greater than that of (1.9). Whatever may be the complete differential operator $\mathcal{D}$ on the LHS of (1.8), the integrand of the Fourier representation of the homogeneous GF must be proportional to $\delta(\tilde{\mathcal{D}}(k))$, since operation on this GF by $\mathcal{D}$ will generate a factor of $\tilde{\mathcal{D}}(k)$ under the integrand; and $q \cdot \delta(q) = 0$, for any quantity q. The constant of proportionality multiplying $\delta(\tilde{\mathcal{D}}(k))$ can be a constant (independent of k_0), or proportional to $s(k_0)$, or some linear combination of these two cases. Explicitly, using the relations: $(a \pm i\epsilon)^{-1} = P(a)^{-1} \mp i\pi\delta(a)$, where P denotes the principal value of the integrand, it is easy to see that the difference of $\tilde{G}_c$ and $\tilde{G}_{ac}$ is proportional to the first of these choices, while the difference of G_R and $\tilde{G}_A$ is proportional to the second.

(iv) Quantum field theory: The difference between the differential operators of (1.9) and (1.10) is simply the inclusion of an additional Laplacian operator in the latter, together with a trivial relabeling of variables. The corresponding GF solutions of the DE

$$(\mu^2 - \partial^2)G(x - x') = \delta(x - x') = \delta(\mathbf{r} - \mathbf{r}')\delta(t - t') \qquad (1.38)$$

may be constructed exactly as in (iii), if one appends a three-dimensional Fourier spatial integration to the latter's integral over k_0. Exactly the same $\pm i\epsilon$ and

$\pm i\epsilon s(k_0)$ factors may be inserted in the inverse of the transform of G, and the same form of results is obtained: four independent solutions to the inhomogeneous DE of (1.38), and two independent solutions to the homogeneous problem. In this case it is most useful to exponentiate the complete denominator of the inhomogeneous GFs, and to use an exponential representation of the $\delta(\tilde{\mathcal{D}}(k))$ of the homogeneous solutions, which steps lead to integrable Gaussian integrals over 4-momenta, and automatically retain the manifest Lorentz invariance of these functions.

That GF bearing the subscript c is called the "causal propagator", and gives the probability amplitude of finding a particle of mass μ at the space–time point x, if it was known to be at the space–time point y; this G_c plays a central role in all QFT calculations. We calculate this Lorentz-invariant function by writing

$$G_c(z) = (2\pi)^{-4} \int d^4k [k^2 + \mu^2 - i\epsilon]^{-1} \exp[ik \cdot z], \qquad (1.39)$$

where $z = x - y$, $d^4k = d^3k\, dk_0$, $k^2 = \mathbf{k}^2 - k_0^2$, and $k \cdot z = \mathbf{k} \cdot \mathbf{z} - k_0 z_0$. An exponential representation for the denominator converts this into

$$G_c = i(2\pi)^{-4} \int_0^\infty ds\, e^{-is\mu^2} \cdot \int d^4k \exp[-isk^2 + ik \cdot z], \qquad (1.40)$$

and the Gaussian k-integrals may be done immediately, remembering the change of sign of the spatial and temporal components of k^2,

$$G_c(z) = (16\pi^2)^{-1} \int_0^\infty ds \cdot s^{-2} \exp[-is\mu^2 + iz^2/4s]. \qquad (1.41)$$

Equation (1.41) is one of the possible representations for Hankel/Bessel functions of order unity, but the specific function depends on whether one is inside, or on, or outside the light cone defined by $z^2 = 0$. The results are

$$G_c(z) = i\mu \frac{\theta(z^2)}{4\pi^2\sqrt{z^2}} K_1(\mu\sqrt{z^2}) - \mu \frac{\theta(-z^2)}{8\pi\sqrt{-z^2}} H_1^{(2)}(\mu\sqrt{-z^2}) + \delta(z^2)/4\pi,$$

$$(1.42)$$

where the last RHS term of (1.42) is that part of the GF appropriate for the light cone. This singular contribution may be obtained by noting that in the limit $z^2 \to 0$, both non-light cone contributions of (1.42) reduce to $(i/4\pi^2)[z^2]^{-1}$, and are independent of μ. One can therefore set $\mu = 0$ in (1.41), and perform the resulting, simple integration to obtain

$$G_c(z) = \frac{i}{4\pi^2} \frac{i}{z^2 + i\epsilon}, \qquad (1.43)$$

the light cone part of which has been written separately in (1.42). With Lorentz indices appended, (1.43) is the ("Feynman gauge") free-photon propagator of

QED, a GF whose configuration-space form, proportional to $[z^2 + i\epsilon]^{-1}$, has the essential simplicity of its momentum-space representation, proportional to $[k^2 - i\epsilon]^{-1}$.

It is interesting to note the way in which causality appears in G_c. Outside the light cone, the K_1 function falls off exponentially, so that relevant signals only propagate on or inside the light cone; and for the latter case, H_1 displays a slow, polynomial fall-off upon which is superimposed oscillatory behavior. Inspection of the intermediate (1.41) shows that if the sign of μ^2 is reversed, then these regions of propagation and non-propagation are effectively interchanged, and one will find a violation of physical causality; this is the so-called "tachyon" situation, which contains propagation outside the light cone, and is clearly unphysical.

1.3 Ordered exponentials

In subsequent representations, ordered exponentials (OEs) appear in a natural and essential way, and it will be useful to set out their general properties at the very beginning. Consider the first-order DE-plus-initial condition

$$dF/d\xi = G(\xi)F(\xi), \ F(0) = 1,$$

whose solution is an ordinary exponential only if $[G(\xi_1), G(\xi_2)] = 0$, for any $0 < \xi_{1,2} < \xi$; here, ξ denotes any relevant variable, such as time or proper time, which appears in the course of analysis. If this commutator does not vanish, then the solution is an OE,

$$F(\xi) = \left(\exp \int_0^\xi d\xi' G(\xi') \right)_+, \qquad (1.44)$$

with the ordering symbol referring to the dummy ξ' variables, where in any expansion of the exponential inside the ordered brackets, the terms are rearranged so that those $G(\xi')$ with the largest values of ξ' stand to the left. For example, the $n = 2$ term is

$$\frac{1}{2!} \int_0^\xi d\xi_1 \int_0^\xi d\xi_2 [G(\xi_1)G(\xi_2)\theta(\xi_1 - \xi_2) + G(\xi_2)G(\xi_1)\theta(\xi_2 - \xi_1)],$$

and both of its contributions are obviously the same, generating

$$\int_0^\xi d\xi_1 \int_0^{\xi_1} d\xi_2 G(\xi_1)G(\xi_2).$$

In the same way, the $(n!)^{-1}$ factor of the nth perturbative term will be removed by its $n!$ equivalent permutations, generating

$$\int_0^\xi d\xi_1 \cdots \int_0^{\xi_{n-1}} d\xi_n G(\xi_1) \cdots G(\xi_{n-1}) G(\xi_n).$$

That this is indeed the proper solution may be verified by differentiation. One calculates a derivative in the ordinary way, "bringing down" the result of differentiation, and placing it anywhere inside the ordered bracket; it is the ordering symbol which states precisely how the terms are to be arranged upon subsequent expansion. One may therefore write

$$\frac{d}{d\xi}\left(\exp \int_0^\xi d\xi' G(\xi')\right)_+ \quad \text{as} \quad \left(G(\xi)\left[\exp \int_0^\xi d\xi' G(\xi')\right]\right)_+,$$

or as

$$G(\xi) \cdot \left(\exp \int_0^\xi d\xi' G(\xi')\right)_+.$$

Upon expansion of the exponent, every $\xi' < \xi$, and hence the $G(\xi)$ factor must always be placed on the extreme LHS, and this OE is therefore the solution of the original DE.

An alternative way to obtain the perturbative expansion of the OE is to rewrite the DE-plus-initial condition of (1.44) as the integral equation

$$F(\xi) = 1 + \int_0^\xi d\xi' G(\xi') F(\xi'), \tag{1.45}$$

and then to expand in powers of G. Simple differentiation of each term, followed by their resummation, shows that one has in this way constructed a solution of the original (1.44).

Just as the solution to that DE is not simply an ordinary exponential, so the derivative of an exponential is not simply given by the product of derivative times exponential,

$$\frac{d}{d\xi} \exp H(\xi) \neq \frac{dH}{d\xi} \cdot \exp H(\xi).$$

The proper definition may be obtained by introducing the quantity

$$Q(\lambda) = \exp[-\lambda H] \cdot \exp[\lambda(H + \delta H)], \qquad Q(0) = 1,$$

and calculating its derivative with respect to λ,

$$\frac{dQ}{d\lambda} = \exp[-\lambda H] \cdot \delta H \cdot \exp[+\lambda H] \cdot Q(\lambda).$$

Hence, to first order in δH,

$$Q(\lambda) \rightarrow 1 + \int_0^\lambda d\lambda' \exp[-\lambda' H] \cdot \delta H \cdot \exp[\lambda' H],$$

and a comparison with the definition of $Q(1)$ yields

$$\frac{d}{d\xi} \exp H(\xi) = \int_0^1 d\lambda \exp[(1-\lambda)H(\xi)] \cdot \frac{dH}{d\xi} \cdot \exp[\lambda H(\xi)]$$

$$= \int_0^1 d\lambda \exp[\lambda H(\xi)] \cdot \frac{dH}{d\xi} \cdot \exp[(1-\lambda)H(\xi)],$$

a most useful and general relation, which reduces to the familiar form only if $[H, dH/d\xi] = 0$. If $H = \int_0^\xi d\xi' G(\xi')$, then

$$\frac{d}{d\xi} \exp\left[\int_0^\xi d\xi' G(\xi')\right]$$

$$= \int_0^1 d\lambda \exp\left[(1-\lambda)\int_0^\xi d\xi' G\right] \cdot G(\xi) \cdot \exp\left[\lambda \int_0^\xi d\xi' G\right]$$

and therefore, upon ordering the exponential,

$$\frac{d}{d\xi}\left(\exp\int_0^\xi d\xi' G(\xi')\right)_+ = G(\xi)\left(\exp\int_0^\xi d\xi' G(\xi')\right)_+,$$

as stated above.

It will also be useful to discuss the functional derivative of an OE, for which it will be convenient to represent $G(\xi)$ as the product $g(\xi)\, C(\xi)$, where $g(\xi)$ is a commuting function, while $[C(\xi_1), C(\xi_2)] \neq 0$. (For example, if $G = i\boldsymbol{\sigma} \cdot \mathbf{B}$, corresponding to a spin interacting with a prescribed magnetic field, the components $\mathbf{B}_i$ are commuting, while the Pauli matrices σ_i do not commute with each other.) Then, if

$$F(\xi, 0) \equiv \left(\exp\left[\int_0^\xi d\xi' g(\xi')C(\xi')\right]\right)_+,$$

one may write

$$\frac{\delta F}{\delta g(\xi)} = \left(C(\xi_1)\exp\left[\int_0^\xi d\xi' g(\xi')C(\xi')\right]\right)_+$$

or

$$\frac{\delta F}{\delta g(\xi_1)} = \left(\exp\left[\int_{\xi_1}^\xi d\xi' g C\right]\right)_+ C(\xi_1)\left(\exp\left[\int_0^{\xi_1} d\xi' g C\right]\right)_+.$$

In other words,

$$\frac{\delta F}{\delta g(\xi_1)} = F(\xi, \xi_1)C(\xi_1)F(\xi_1, 0), \qquad (1.46)$$

in which the correct ordering of all non-commuting factors is maintained.

How does one write non-perturbative expressions for OEs? In general, this is an open and quite important question, and very little is known. In Chapter 9 we describe relatively simple "adiabatic" and "stochastic" non-perturbative approximations for a unitary OE; but between these limits there is a wide range of interesting behavior which still awaits proper classification and description. Quite generally, OEs are at the heart of Fradkin-GF representations in QED and QCD; and it is the lack of adequate, non-perturbative methods for estimating relevant OEs which has, in part, forced so much of the interesting Physics of these fields to be described in perturbative terms.

Notes

1 A miller's son, born in the England of two centuries ago, Green was trained as a baker and was essentially self-taught in Mathematical Physics. After publishing in 1828 the Theorem which bears his name (and coining the word "potential" in electrostatics), his mathematical talent and physical insight were eventually recognized, and at the age of 40 he entered Caius College, Cambridge, as a student. He graduated with honors in 1837, and was appointed a Fellow of Caius College; four years later he died. His collected works, brought in part to the attention of the scientific community by Kelvin, were edited by N. M. Ferrers and first published under the title *Mathematical Papers of George Green*, by MacMillan & Co., London, 1871; they were republished by the Chelsea Publishing Co., Bronx, NY, in 1970. Reading these Papers, one is struck by the fresh and modern exposition. Vector analysis had not yet been invented, and so his was the language of individual components; but Green's physical arguments were virtually identical to those in use today.
2 One excludes in the integration over $\mathbf{r}'$ a small region of radius ϵ centered about the point $\mathbf{r}$, applies Gauss' theorem to the entire but restricted volume of integration, and then takes the limit $\epsilon \rightarrow 0$. One of the most rewarding treatments of this *genre* in electromagnetism, fluids, and even special relativity, which makes use of modern vector notation and vector calculus – and when written was accompanied by a plea for the more widespread use of vectorial methods – may be found in the small but unforgettable book *Advanced Vector Analysis* by the Australian physicist C. E. Weatherburn, published by G. Bell & Sons, London, 1924, 1928, 1937, 1943.
3 E. S. Fradkin, *Nucl. Phys.* **76** (1966) 588. The detailed exposition of Fradkin's representation found in Chapter 3 follows the presentation previously given in HMF#2.
4 See, for example, HMF#2, Chapter 15.

2

Elementary functional methods

It is the existence and extraordinary usefulness of the functional representations for GFs, introduced by Schwinger, Symanzik and Fradkin, by Feynman in terms of path integrals, and discussed at length in the next chapter, that forces a brief discussion of functional methods here.[1] In certain subjects, such as QFT, the conceptual simplicity permitted by a functional description is remarkable, allowing one to view the manner in which different correlation functions are related to each other within a basic, functional formalism, either by a specific choice of interaction, or by restrictions following from unitarity.[2] To learn functional methods for the first time is rather like being given the opportunity to "see the forest, instead of individual trees", for one is then made aware of the existence of a vast overview of the nonlinear problem at hand.

The intention of this chapter is to provide, mainly for those readers who have never been familiar with the subject, a simple and straightforward explanation of functional methods, along with relevant short cuts and tricks of the functional trade. One or another aspects of these manipulations will then be freely employed in subsequent chapters.

2.1 Functional differentiation

This simplest generalization of ordinary differentiation may be defined as follows. Suppose one has a *functional* of $j(x)$, that is, dependence on j which can be represented as the sum of many powers of j, each multiplied by an appropriate weighting function and integrated over all coordinates,

$$F[j] = F_0 + \int F_1(u)j(u) + \frac{1}{2!} \int F_2(u_1, u_2)j(u_1)j(u_2)$$

$$+ \frac{1}{3!} \int F_3(u_1, u_2, u_3)j(u_1)j(u_2)j(u_3) + \cdots. \quad (2.1)$$

17

Here, F_0 is a constant, and the $F_n(u_1, \ldots, u_n)$ are specified, symmetric functions of their variables, with integration over the entire (finite or infinite) range of all variables. This supposes that $F[j]$ has a Taylor expansion in powers of j, a simplification not necessary for the general definition, but one that is more readily grasped and discussed. It is in this sense that $F[j]$ is understood to be a "functional of j". Unless otherwise noted, $\int fj$ will be used to represent an integration over all, relevant (for example, n-dimensional space time) variables, $\int d^n u f(u) j(u)$.

Functional differentiation may now be defined in a manner paralleling that of ordinary differentiation. Suppose we write explicitly the coordinate u of $j(u)$ in $F[j]$; that is, u is to denote any one of the variables on the RHS of (2.1). Then, one defines

$$\frac{\delta F}{\delta j(x)} = \lim_{\epsilon \to 0} \cdot \frac{1}{\epsilon} \{F[j(u) + \epsilon \delta(x - u)] - F[j(u)]\}. \qquad (2.2)$$

From (2.2), there follow the simple examples

$$\frac{\delta}{\delta j(x)} \exp\left[\int fj\right] = f(x) \exp\left[\int fj\right],$$

and

$$\frac{\delta}{\delta j(x)} \exp\left[\frac{i}{2} \int j(u) f(u, v) j(v)\right] = i \int f(x, z) j(z) \cdot \exp\left[\frac{i}{2} \int jfj\right],$$

etc.

2.2 Linear translation

There is one type of functional operation, involving an infinite number of functional derivatives, which appears frequently and which can be understood in complete analogy to the similar translation operation of the ordinary calculus. That is, if

$$\exp\left[a\frac{d}{dx}\right] \cdot f(x) = f(x + a)$$

represents the ordinary translation (given by the Taylor expansion in powers of a), the functional translation of $F[j]$ is accomplished by

$$\exp\left[\int f\frac{\delta}{\delta j}\right] \cdot F[j] = F[j + f], \qquad (2.3)$$

where $\int f\frac{\delta}{\delta j} = \int d^n u f(u) \frac{\delta}{\delta j(u)}$.

Equation (2.3) is surely intuitive, and may be easily proven by replacing f on the LHS of (2.3) by λf, and constructing the DE corresponding to variations of the parameter λ. If

$$F_\lambda[j] \equiv \exp\left[\lambda \int f \frac{\delta}{\delta j}\right] \cdot F[j],$$

then

$$\frac{\partial F_\lambda}{\partial \lambda}[j] = \int f(u) \frac{\delta}{\delta j(u)} F_\lambda[j]. \tag{2.4}$$

The general solution to (2.4) can be found in a variety of ways, but we do this here by a method which will have application to other, more difficult problems. One first finds a convenient representation for an arbitrary $F[j]$, which can be given in terms of functional differentiation with respect to a different source function, $g(z)$; that is, for the functional F of (2.1), one may write

$$F[j] = \left\{ F_0 + \int F_1(u) + \frac{1}{i} \frac{\delta}{\delta g(u)} \right. $$
$$\left. + \frac{1}{2!} \int F_2(u_1, u_2) \frac{1}{i} \frac{\delta}{\delta g(u_1)} \cdot \frac{1}{i} \frac{\delta}{\delta g(u_2)} + \cdots \right\} \cdot \exp\left[i \int gj\right]\bigg|_{g \to 0},$$

or

$$F[j] = F\left[\frac{1}{i}\frac{\delta}{\delta g}\right] \cdot \exp\left[i \int gj\right]\bigg|_{g \to 0}, \tag{2.5}$$

where one is instructed, in (2.5), to set the source g equal to zero after all necessary derivatives with repect to g have been taken. This is convenient for our purposes, for all the functions $F_n(u_1, \ldots, u_n)$ are combined with functional derivatives with repect to g to form the operator $F(\delta/i\delta g)$; all the j-dependence sits in the exponential factor of (2.5) and commutes with the $\delta/\delta g$ operations. If, therefore, one can solve (2.4) for the special choice $F_0[j] = \exp[i \int gj]$ for arbitrary g, then the application of $F(\delta/i\delta g)$ to that F will produce a solution of (2.4) for a general $F[j]$.

Accordingly, we rewrite (2.4) for the special functional $F_0[j] = \exp[i \int gj]$, $F_\lambda^0[j] = \exp[\lambda \int f \frac{\delta}{\delta j}] \cdot F_0[j]$, which now, clearly, satisfies the DE

$$\frac{\partial}{\partial \lambda} F_\lambda^0[j] = \left(i \int fg\right) F_\lambda^0[j]. \tag{2.6}$$

With the proper boundary condition at $\lambda = 0$, the solution to (2.6) is immediate,

$$F_\lambda^0[j] = \exp\left[i\lambda \int fg\right] \cdot F_0[j] = F_0[j + \lambda f]. \tag{2.7}$$

Operation upon both sides of (2.7) with $F(\delta/i\delta g)$, for $\lambda = 1$, then yields the result (2.3) for a general functional. An equivalent but somewhat simpler derivation may be obtained by using the idea of a functional Fourier transform, as in the discussion following (2.18), below.

2.3 Quadratic (Gaussian) translation

There is another form of "translation" which has relevance in another functional context, corresponding to the use of a "quadratic translation" operator,

$$\exp \mathcal{D} = \exp\left[-\frac{i}{2} \int \frac{\delta}{\delta j(u)} A(u, v) \frac{\delta}{\delta j(v)} \right],$$

where $A(u, v)$ is a symmetric function of its variables; and we next consider its application upon some simple forms. It turns out that $[\exp \mathcal{D}]F[j]$ corresponds to a functional integral (FI) of Gaussian-weight upon F, and may be – with a few exceptions – be obtained in closed form only when $F[j]$ itself is not more complicated than a Gaussian.

The simplest quantity of this form is $\exp \mathcal{D} \cdot \exp[i \int gj]$, and may be evaluated by replacing A, in $\mathcal{D}$, by λA, and then obtaining and solving the simple DE corresponding to variation of λ. One immediately finds

$$\exp\left[-\frac{i}{2} \int \frac{\delta}{\delta j} A \frac{\delta}{\delta j} \right] \cdot \exp\left[i \int jg \right] = \exp\left[\frac{i}{2} \int gAg + i \int jg \right]. \quad (2.8)$$

There is a simple but useful generalization of (2.8) which involves the action of $\exp \mathcal{D}$ upon a product of functionals, say $F_1[j]$ and $F_2[j]$. Again, as in (2.5), it is simplest to write each of these as $F_i(\delta/i\delta g_i) \cdot \exp[i \int g_i j]$, with the g_i vanishing after all derivatives are taken. Then, since $\exp \mathcal{D}$ commutes with the $F(\delta/i\delta g_i)$, one can write

$$\exp\left[-\frac{i}{2} \int \frac{\delta}{\delta j} A \frac{\delta}{\delta j} \right] \cdot F_1[j] \cdot F_2[j]$$

$$= F_1\left[\frac{1}{i} \frac{\delta}{\delta g_1} \right] \cdot F_2\left[\frac{1}{i} \frac{\delta}{\delta g_2} \right] \cdot \exp\left[-\frac{i}{2} \int \frac{\delta}{\delta j} A \frac{\delta}{\delta j} \right] \cdot \exp\left[i \int j(g_1 + g_2) \right]\Bigg|_{g_i \to 0},$$

which, using (2.8), can be rewritten as

$$F_1 \cdot F_2 \cdot \exp\left[\frac{i}{2} \int g_1 A g_1 + \frac{i}{2} \int g_2 A g_2 + i \int g_1 A g_2 \right]$$

$$\cdot \exp\left[i \int j(g_1 + g_2) \right]\Bigg|_{g_i \to 0},$$

or, with (2.3) and $\mathcal{D}_i = -\frac{i}{2} \int \frac{\delta}{\delta J_i} A \frac{\delta}{\delta J_i}$, $\mathcal{D}_{12} = -i \int \frac{\delta}{\delta j_1} \cdot A \cdot \frac{\delta}{\delta j_2}$, can be written in the form

$$e^{\mathcal{D}} \cdot F_1[j]F_2[j] = e^{\mathcal{D}_{12}}[(e^{\mathcal{D}_1} F_1[j_1])(e^{\mathcal{D}_2} F_2[j_2])]|_{j_1=j_2=j}, \qquad (2.9)$$

and can be extended to products of more than two functionals in an obvious manner. In words: $\exp \mathcal{D}$ is a "linkage operator", which first links all pairs of j-factors within each functional by the operation of $\exp \mathcal{D}$ on that functional (these may be called the "self-linkages"); and which then links the different functionals by means of the operators $\exp[\mathcal{D}_{ij}]$. These forms will appear with a certain frequency when describing the Green's functions of QFT.

The evaluation of $\exp \mathcal{D}$ upon the Gaussian functional $\exp[\frac{i}{2} \int jBj]$ is somewhat more interesting, and we begin by following the parametric method of Zumino[3] and Sommerfield,[4] by considering the quantity

$$F_\lambda[j] = \exp\left[-\frac{i}{2}\lambda \int \frac{\delta}{\delta j} A \frac{\delta}{\delta j}\right] \cdot \exp\left[\frac{i}{2} \int jBj\right], \qquad (2.10)$$

where both $A(u, v)$ and $B(u, v)$ are symmetric functions of their arguments. Again, one constructs the DE corresponding to variation of λ,

$$\frac{\partial F_\lambda[j]}{\partial \lambda} = -\frac{i}{2} \int \frac{\delta}{\delta j(u)} A(u, v) \frac{\delta}{\delta j(v)} \cdot F_\lambda[j]. \qquad (2.11)$$

An intuitive ansatz is then chosen for this Gaussian quantity,

$$F_\lambda[j] = \exp\left[\frac{i}{2} \int j\chi_\lambda j + i \int h_\lambda j + L_\lambda\right], \qquad (2.12)$$

where $\chi_\lambda(u, v)$, $h_\lambda(u)$, and L_λ are three functions of u, v, and λ to be determined. Note that these quantities satisfy the boundary conditions: $\chi_0(u, v) = B(u, v)$, $h_0(u) = 0$, $L_0 = 0$.

Substituting (2.12) into (2.11), canceling a factor of $F_\lambda[j]$ from both sides of the equation, and equating coefficients of $j(z)$ of the remaining terms, one finds the simultaneous relations

$$\frac{d\chi_\lambda}{d\lambda}(u, v) = \int \chi_\lambda(u, w)A(w, z)\chi_\lambda(z, v),$$

$$\frac{dh_\lambda}{d\lambda}(u) = \int \chi_\lambda(u, w)A(w, z)h_\lambda(z),$$

and

$$\frac{dL_\lambda}{d\lambda} = \frac{1}{2} \int A(u, v)\chi_\lambda(v, u) + \frac{i}{2} \int h_\lambda(u)A(u, v)h_\lambda(v).$$

In a more compact, matrix notation, these equations read

$$\frac{d\chi_\lambda}{d\lambda} = \chi_\lambda A \chi_\lambda, \qquad \frac{dh_\lambda}{d\lambda} = \chi_\lambda A h_\lambda, \qquad \frac{dL_\lambda}{d\lambda} = \frac{1}{2}\text{Tr}[A\chi_\lambda] + \frac{i}{2}h_\lambda^T A h_\lambda,$$

and can be solved making use of the conditions: $\chi_0 = B$, $h_0 = 0$, $L_0 = 0$. One immediately sees that $h_\lambda(u) = 0$; that $\chi_\lambda = B[1 - \lambda AB]^{-1}$; and that $L_\lambda = -(1/2)\text{Tr}\ln[1 - \lambda AB]$. With $\lambda = 1$, we then have the desired formula,

$$\exp\left[-\frac{i}{2}\int \frac{\delta}{\delta j} A \frac{\delta}{\delta j}\right] \cdot \exp\left[\frac{i}{2}\int jBj\right]$$

$$= \exp\left[\frac{i}{2}\int jB(1 - AB)^{-1}j - \frac{1}{2}\text{Tr}\ln(1 - AB)\right]. \qquad (2.13)$$

For this shorthand, matrix notation, $\langle x|A|y\rangle = A(x, y)$, $\langle x|B|y\rangle = B(x, y)$, while the matrix element $\langle x|\bar{B}_\lambda|y\rangle = \langle x|B[1 - \lambda AB]^{-1}|y\rangle = \bar{B}_\lambda(x, y)$ is to satisfy the integral equations

$$\bar{B}_\lambda(x, y) = B(x, y) + \lambda \int B(x, u)A(u, v)\bar{B}_\lambda(v, y),$$

or

$$\bar{B}_\lambda(x, y) = B(x, y) + \lambda \int \bar{B}_\lambda(x, u)A(u, v)B(v, y).$$

In terms of this function, the formal expression above for L_λ is simply the statement: $L_\lambda = (1/2)\int_0^\lambda d\lambda' \int A(u, v)\bar{B}_{\lambda'}(v, u)$. It will shortly become clear that (2.13) is a statement of Gaussian functional integration.

An immediate and most useful generalization of (2.13) is

$$\exp\left[-\frac{i}{2}\int \frac{\delta}{\delta j} A \frac{\delta}{\delta j}\right] \cdot \exp\left[\frac{i}{2}\int jBj + i\int fj\right]$$

$$= \exp\left[\frac{i}{2}\int jB(1 - AB)^{-1}j + i\int j(1 - BA)^{-1}f\right.$$

$$\left. +\frac{i}{2}\int fA(1 - BA)^{-1}f - \frac{1}{2}\text{Tr}\ln(1 - AB)\right]. \qquad (2.14)$$

In this case, the $h_\lambda(u)$ term of the ansatz (2.12) turns out to be non-zero and easily calculable.

A slightly different form of (2.13) is useful in charged-boson situations,

$$\exp\left[-i\int \frac{\delta}{\delta j} A \frac{\delta}{\delta j^*}\right] \cdot \exp\left[i\int j^* Bj\right]$$

$$= \exp\left[i\int j^* B(1 - AB)^{-1}j - \text{Tr}\ln(1 - AB)\right], \qquad (2.15)$$

and is easily obtained by the same parametric technique. Note that $A(u, v)$ and $B(u, v)$ are here no longer symmetric functions of their variables.

There exists a simple extension of (2.13) which is of immediate use in theories containing fermions. If $A_{\alpha\beta}(u, v)$ and $B_{\alpha\beta}(u, v)$ are now non-symmetric, (Dirac) matrix-valued functions, and if the c-number fermionic sources $\eta_\alpha(u)$, $\bar{\eta}_\beta(v)$ which replace the bosonic sources $j(u)$ are taken to be elements of the anti-commuting (Grassmann) algebra, satisfying

$$\{\eta_\alpha(u), \eta_\beta(v)\} = \{\eta_\alpha(u), \bar{\eta}_\beta(v)\} = 0,$$

$$\left\{\frac{\delta}{\delta\eta_\alpha(u)}, \eta_\beta(v)\right\} = \left\{\frac{\delta}{\delta\bar{\eta}_\alpha(u)}, \bar{\eta}_\beta(v)\right\} = \delta_{\alpha\beta}\delta(u - v),$$

then a very similar construction may be carried through, yielding

$$\exp\left[-i\int\frac{\delta}{\delta\eta_\alpha}A_{\alpha\beta}\frac{\delta}{\delta\bar{\eta}_\beta}\right]\cdot\exp\left[i\int\bar{\eta}_\alpha B_{\alpha\delta}\eta_\delta\right]$$

$$= \exp\left[i\int\bar{\eta}B(1 + AB)^{-1}\eta + \mathrm{Tr}\ln(1 + AB)\right]. \qquad (2.16)$$

Dirac matrices have been suppressed on the RHS of (2.16), but it should be noted that the trace operation Tr here includes a summation over Dirac indices as well as over space–time coordinates.

Note also that factors of $(1/2)$ are absent, and that the sign of A on the RHS of (2.16) appears to be reversed in comparison with the bosonic result (2.13). More importantly, the sign of $\mathrm{Tr}\ln[1 + AB]$ is reversed, which is the origin of the "extra minus sign multiplying each fermionic closed-loop" rule of perturbation theory, and which plays a role in the "supersymmetric" cancellations of divergences between bosonic and fermionic closed-loops.[5]

2.4 Functional integration

The object here is to generalize ordinary integration over a real variable, $\int dx$, to integration over a real function, $u(x)$. One must first define the integration measure, which is typically done by breaking up all space–time into a fine mesh of N small cells of volume Δ, each labeled by a subscript i referring to the ith cell: $u(x) \rightarrow u(x_i) = u_i$. Then, the functional integral (FI) $\int d[u]$ is defined as the product over all ordinary integrals of the u_i,

$$\int d[u] = \lim_{N\to\infty}\prod_{i=1}^{N}\int_{-\infty}^{+\infty}du_i. \qquad (2.17)$$

Sometimes a divergent normalization factor is included in this definition; but it is not really necessary as long as proper care is taken to normalize all physical expressions.

The simplest, non-trivial integrand to insert under the FI is one which leads to a Dirac δ-functional, expressing the equality of one function with another for all values of their arguments. Following the spirit of (2.17), one writes $\int dx u(x)[j(x) - f(x)]$ as $\Delta \sum_{i=1}^{N} u_i(j_i - f_i)$, and calculates

$$\delta[j - f] = \eta^{-1} \int d[u] \exp\left[i \int u(j - f)\right], \qquad (2.18)$$

with η a normalization constant to be determined. Equation (2.18) represents the δ-functional, defined in this way as the product of N ordinary δ-functions, one at each space–time coordinate i,

$$\eta^{-1} \prod_{i=1}^{N} \int_{-\infty}^{+\infty} du_i \exp\left[i\Delta u_i(j_i - f_i)\right] = \eta^{-1} \left(\frac{2\pi}{\Delta}\right)^N \prod_{i=1}^{N} \delta(j_i - f_i).$$

With the normalization η chosen as $(2\pi/\Delta)^N$, this $\delta[j - f]$ will under subsequent functional integration act to replace each $j(x)$ by the function $f(x)$.

Equation (2.18) is a special case of a functional Fourier transform (FFT), in the sense that the FFT of $\delta[u]$ is a constant. It is frequently useful to imagine an arbitrary functional $F[j]$ as given by its FFT,

$$F[j] = \eta^{-1} \int d[u] \tilde{F}[u] \exp\left[i \int ju\right],$$

and we have already done the equivalent thing in constructing the solution to (2.4), above. Again breaking up the space–time region into a fine mesh of small volumes Δ, the existence of the FFT can be understood in terms of the existence of an ordinary FT at each mesh coordinate.

It may also be noted that the measure of the FI can be defined in terms of the FT $\tilde{u}(k)$ of $u(x)$, by breaking up all of k-space into a fine mesh, and integrating the $\tilde{u}(k)$ in each mesh volume. However, if $u(x)$ is real, $\tilde{u}(k)$ is complex, and one must be careful to integrate over both the real and imaginary parts of each mesh variable $\tilde{u}_i = \tilde{u}(k_i)$.

Just as for the case of ordinary integration, the most complicated FI that can be performed exactly is a Gaussian. One requires

$$I[j; A] = \int d[u] \exp\left[\frac{i}{2} \int uAu + i \int ju\right], \qquad (2.19)$$

where, again, $A(x, y)$ is a symmetric function of its variables, and $j(x)$ is an arbitrary source function. Without performing any calculation at all one can obtain the j-dependence of (2.19), by making the variable change: $u(x) = v(x) - \int dz A^{-1}(x, z) j(z)$, where the quantity A^{-1} is assumed to exist, and to satisfy the relations

$$\int dz A(x, z) A^{-1}(z, y) = \int dz A^{-1}(x, z) A(z, y) = \delta(x - y).$$

The exponential of (2.17) then becomes

$$\frac{i}{2} \int v A v - \frac{i}{2} \int j(x) A^{-1}(x, y) j(y),$$

so that $I[j; A] \sim \exp[-(i/2) \int j A^{-1} j]$. The proportionality constant, and in particular its A-dependence, is a quantity of considerable interest, and the most straightforward way to obtain it is by direct integration.

Imagine an orthogonal matrix $\langle x | M | y \rangle = M(x, y)$ with a continuous number of indices, satisfying the normalization condition

$$\int dz M(x, z) M^{\mathrm{T}}(z, y) = \delta(x - y), \tag{2.20}$$

with $M^{\mathrm{T}}(x, y) = M(y, x) = M^{-1}(x, y)$. When configuration space is broken up into small cells of 4-volume Δ, there will be correspondingly many discrete components of this matrix, M_{ij}; and with the replacement $\delta(x - y) \to \delta_{ij}/\Delta$, one finds the discrete version of (2.20),

$$\sum_{l} M_{il} M_{lj}^{\mathrm{T}} = \delta_{ij}/\Delta^2. \tag{2.21}$$

The orthogonal M is to be chosen such that it diagonalizes A,

$$\int M^{\mathrm{T}}(x, z) A(z, w) M(w, y) = \delta(x - y) a(x), \tag{2.22}$$

where $a(x)$ represents the continuous-valued eigenvalue of A. [Note that for $A(x, y) = \delta(x - y)$, (2.22) reduces to (2.20) if $a = 1$.] For discrete indices (2.22) becomes

$$\sum_{\alpha\beta} M_{i\alpha}^{\mathrm{T}} A_{\alpha\beta} M_{\beta j} = \delta_{ij} a_i / \Delta^3. \tag{2.23}$$

Indeed, for any $f(A)$ expressible as an infinite polynomial in A, these become

$$\int M^{\mathrm{T}}(x, z) \langle z | f(A) | w \rangle M(w, y) = \delta(x - y) f(a(x)), \tag{2.24}$$

and

$$\sum_{\alpha\beta} M_{i\alpha}^{\mathrm{T}} f_{\alpha\beta}(A) M_{\beta j} = \delta_{ij} f(a_i)/\Delta^3, \tag{2.25}$$

where $\langle x|f(A)|y\rangle$ and f_{ij} denote projections of $f(A)$ in the continuous and discrete cases, respectively.

Under the variable change: $u(x) = \int dz M(x,z)q(z)$, the FI of (2.19) becomes

$$I[j;A] = \int d[g] \exp\left[\frac{i}{2}\int g^2(x)a(x) + i\int g(x)J(x)\right], \quad (2.26)$$

where the "new source" $J(x)$ is related to $j(x)$ by: $J(x) = \int M^{\mathrm{T}}(x,z)j(z)$. Note that (2.21) implies that the new, discrete measure $d[g]$ is exactly the same as the original $d[u]$, since the $N \times N$ matrix ΔM_{ij} is orthogonal, with determinant unity. The discrete version of (2.26) is

$$\prod_{i=1}^{N}\int_{-\infty}^{+\infty} dg_i \exp\left[\frac{i}{2}\Delta a_i g_i^2 + i\Delta g_i J_i\right] = \prod_{i=1}^{N}\left(\frac{2\pi i}{\Delta a_i}\right)^{1/2} e^{-\frac{i}{2}\Delta J_i^2/a_i}. \quad (2.27)$$

With the $f(A)$ of (2.24) and (2.25) given by $f(A) = A^{-1}$, the product over all cells of the exponential factor on the RHS of (2.27) yields

$$\exp\left[-\frac{i}{2}\int j(x)A^{-1}(x,y)j(y)\right],$$

which is just that found above without any integration. With $f(A)$ now chosen as $f = \ln(A)$, the A-dependence of the factors

$$\prod_{i=1}^{N}(a_i)^{-1/2} = [\det a]^{-1/2} = \exp\left[-\frac{1}{2}\sum_i \ln a_i\right]$$

can be written in terms of: $\int dx \langle x|\ln(A)|x\rangle = \mathrm{Tr}\ln(A)$, after passing to the continuous limit. Denoting the remaining $(2\pi i/\Delta)^{N/2}$ by the constant C, one has the precise evaluation

$$I[j;A] = C \exp\left[-\frac{i}{2}\int jA^{-1}j - \frac{1}{2}\mathrm{Tr}\ln A\right]. \quad (2.28)$$

Every physically-significant computation of an FI such as (2.19) is always phrased in such a way that the divergent, normalization constant C disappears from the final result.

There is another way of evaluating (2.19) which brings out the equivalence of the parametrically-obtained relation (2.13) and Gaussian functional integration. From the definition (2.19) and the translation property (2.3), there follows

$$\exp\left[-\frac{i}{2}\int \frac{\delta}{\delta j} D \frac{\delta}{\delta j}\right] \cdot I[j;A] = I[j;A+D]. \quad (2.29)$$

If one substitutes into (2.29) the ansatz

$$I[j; A] = N[A] \cdot \exp\left[-\frac{i}{2} \int jA^{-1}j\right],\qquad(2.30)$$

there results

$$N[A] \exp\left[-\frac{1}{2}\mathrm{Tr}\ln(1 + DA)^{-1}\right] = N[A + D],\qquad(2.31)$$

where a common factor $\exp[-i/2 \int j(A + D)^{-1}j]$ has canceled from both sides of (2.31). The latter can be rewritten in the form

$$N[A] \exp\left[\frac{1}{2}\mathrm{Tr}\ln A\right] = N[A + D] \exp\left[\frac{1}{2}\mathrm{Tr}\ln(A + D)\right],\qquad(2.32)$$

and since the LHS depends only on A, while the RHS depends only on $A + D$, both sides must be constant, independent of A and D, from which follows

$$N[A] = C \cdot \exp\left[-\frac{1}{2}\mathrm{Tr}\ln A\right].\qquad(2.33)$$

With C chosen as the constant of (2.28), (2.33) and (2.30) again reproduce (2.28).

The corresponding, direct functional integration over fermionic variables is left as an exercise for the interested reader.[6] It is less intuitive than that of the bosonic case because of the presence of anticommuting variables, but it is somewhat simpler because all products of like-Grassmannian variables vanish. Equation (2.16) plays an essential role in all field theories containing fermions, and will find application in the next section.

2.5 Examples drawn from quantum field theory

Illustrations of the functional techniques of the previous sections, which serve to illustrate the ability of functional methods to display an overall view of the entire structure of interactions in realistic QFTs, are presented here by two examples drawn from QED and from a generically related pair of interacting scalar fields. For each case, we begin by writing Schwinger's formal, functional solution for the relevant generating functional; and we shall use units such that $h = c = 1$.

(1) A pair of interacting scalar fields may be described by the Lagrangian density

$$\mathcal{L}^1 = -\frac{1}{2}[\mu^2 A^2 + (\partial A)^2] - \frac{1}{2}[m^2 B^2 + (\partial B)^2] - \frac{g}{2}B^2 A,\qquad(2.34)$$

and corresponds to scalar particles of mass m interacting with scalar particles of mass μ. [The masses and charges in these examples are "bare", unrenormalized quantities. Renormalization comes later on, when one makes the transition, by calculating relevant elements of the S-matrix, from the underlying fields to properties of measurable particles.]

All of the correlation functions, or n-point functions, of this theory may be obtained by functional differentiation of the generating functional ($\hat{G}F$)

$$\mathcal{Z}[j, k] = \langle 0| \left(\exp i \int [jA + kB] \right)_+ |0\rangle$$

with respect to its sources, where $j(z)$ and $k(z)$ are c-number, bosonic sources associated with the boson field operators $A(z)$ and $B(z)$; and where the ordering of this OE is with respect to the time-variable of the 4-dimensional integrals $\int d^4x = \int d^3x \int dt$. Functional differentiation, followed by vanishing sources, then generates the desired correlation functions,

$$\langle 0|(A(x_1)\cdots A(x_n)B(y_1)\cdots B(y_l))_+|0\rangle$$

$$= \frac{1}{i}\frac{\delta}{\delta j(x_1)}\cdots\frac{1}{i}\frac{\delta}{\delta j(x_n)}\cdot\frac{1}{i}\frac{\delta}{\delta k(y_1)}\cdots\frac{1}{i}\frac{\delta}{\delta k(y_l)}\mathcal{Z}\bigg|_{j=k=0}.$$

In this case, the Schwinger/Symanzik constructions described in refs. 1 and 2 of the Preface provide the formal solution

$$\mathcal{Z}[j, k] = \langle S\rangle^{-1} e^{-i\frac{g}{2}\int(\frac{1}{i}\frac{\delta}{\delta k})^2(\frac{1}{i}\frac{\delta}{\delta j})} \cdot e^{\frac{i}{2}\int jD_c j} \cdot e^{\frac{i}{2}\int k\Delta_c k}, \tag{2.35}$$

where $\langle S\rangle$ denotes the relevant vacuum-to-vacuum amplitude, and where D_c and Δ_c are the free-particle propagators of masses μ and m, respectively. Note that the argument of the functional operator of (2.35) is given by the interaction part of the action, with field operators replaced by factors of $-i$ multiplying functional derivatives of their sources; and that this functional operator then acts upon the product of free-particle GFs.

With (2.13), the quantum fluctuations of the B-field may be performed, leading to

$$\mathcal{Z}[j, k] = e^{\frac{i}{2}\int kG_c[\frac{1}{i}\frac{\delta}{\delta j}]k} \cdot \frac{e^{L[\frac{1}{i}\frac{\delta}{\delta j}]}}{\langle S\rangle} \cdot e^{\frac{i}{2}\int jD_c j} \tag{2.36}$$

where $G_c(x, y|A)$ denotes the scalar GF $\langle x|G_c[A]|y\rangle$, $G_c[A] = [m^2 - \partial^2 + gA]^{-1}$, and $L[A] = -\frac{1}{2}\text{Tr}\ln(1 + gA\Delta_c) = -\frac{g}{2}\int_0^1 d\lambda \int d^4x G_c(x, x|\lambda A) \cdot A(x)$.

With the aid of the elementary manipulation

$$F\left[\frac{1}{i}\frac{\delta}{\delta j}\right]e^{\frac{i}{2}\int jD_c j} = e^{\frac{i}{2}\int jD_c j} \cdot e^{-\frac{i}{2}\int\frac{\delta}{\delta A}D_c\frac{\delta}{\delta A}}F[A],$$

which may easily be demonstrated using our previous representation of an arbitrary functional, $F[A] = F(\frac{1}{i}\frac{\delta}{\delta j})\exp[i\int gA]|_{g\to 0}$, (2.36) may be rephrased in a somewhat more intuitive form, as

$$\mathcal{Z}[j,k] = e^{\frac{i}{2}\int jD_c j} \cdot e^{\mathcal{D}_A} \cdot e^{\frac{i}{2}\int kG_c[A]k} \cdot \frac{e^{L[A]}}{\langle S\rangle}\Bigg|_{A\equiv\int D_c j}, \quad (2.37)$$

expressed in terms of the linkage operator $\exp[\mathcal{D}_A]$, where $\mathcal{D}_A = -\frac{i}{2}\int\frac{\delta}{\delta A(x)} \times D_c(x-y)\frac{\delta}{\delta A(y)}$. From the normalization requirement $\mathcal{Z}[0,0] = 1$, it then follows that

$$\langle S\rangle = \exp[\mathcal{D}_A]\cdot\exp(L[A])|_{A\to 0}. \quad (2.38)$$

(2) QED is described by the Lagrangian density

$$\mathcal{L}' = -\bar{\psi}[m + r\cdot\partial]\psi - \frac{1}{4}\sum_{\mu,\nu}F_{\mu\nu}^2 + ig\bar{\psi}\gamma\cdot A\psi,$$

and has the corresponding ĜF

$$\mathcal{Z}[j,\eta,\bar{\eta}] = \langle 0|\left(\exp\left[i\int(\bar{\psi}\eta + \bar{\eta}\psi + j\cdot A)\right]\right)_+|0\rangle.$$

Here, $j_\mu(z)$ is a bosonic c-number, 4-vector source; but the spinorial sources $\eta_\alpha(x)$, $\bar{\eta}_\beta(y)$ are Grassmann variables, anticommuting with themselves and with all fermion fields, as in (2.16).

Schwinger's functional solution for QED may then be written as

$$\mathcal{Z}[j,\eta,\bar{\eta}] = \langle S\rangle^{-1}e^{i\int\frac{\delta}{\delta\eta}(g\gamma\cdot\frac{\delta}{\delta j})\frac{\delta}{\delta\bar{\eta}}}\cdot e^{i\int\bar{\eta}S_c\eta}\cdot e^{\frac{i}{2}\int jD_c j},$$

and the quantum fluctuations of the ψ, $\bar{\psi}$ fields may be calculated using (2.16), and yield

$$\mathcal{Z}[j,\eta,\bar{\eta}] = e^{\frac{i}{2}\int jD_c j}\cdot e^{\mathcal{D}_A}\cdot e^{i\int\bar{\eta}G_c[A]\eta}\cdot\frac{e^{L[A]}}{\langle S\rangle}, \quad (2.39)$$

where $A_\mu(z) = \int d^4w D_{c,\mu\nu}(z-w)j_\nu(w)$, $\mathcal{D}_A = -\frac{i}{2}\int\frac{\delta}{\delta A_\mu}D_{c\mu\nu}\frac{\delta}{\delta A_\nu}$, $G_c[A] = [m + \gamma\cdot(\partial - igA)]^{-1}$, and $L[A] = +\mathrm{Tr}\ln[1 - ig\gamma\cdot AS_c]$. Here, D_c and S_c are free-particle photon and fermion propagators, respectively. Again,

$$\langle S\rangle = \exp[\mathcal{D}_A]\exp\{L[A]\}|_{A\to 0}. \quad (2.40)$$

From these two examples one sees in each case the difference between the "first quantization" of potential theory, and the full, "second quantized" QFT.

In the first case, the B particles have interactions only with a specified, external field $A^{\text{ext}}(z)$, corresponding to the ĜFs

$$\mathcal{Z}[k] = \exp\left[\frac{\mathrm{i}}{2}\int kG_{\text{c}}[A^{\text{ext}}]k\right] \quad \text{or} \quad \mathcal{Z}[\eta, \bar{\eta}] = \exp\left[\mathrm{i}\int \bar{\eta}G_{\text{c}}[A^{\text{ext}}]\eta\right],$$

while in the second case, one requires the functional linkage operator to act upon products of $G_{\text{c}}[A]$ and $L[A]$. For example, the probability amplitude for the scattering of a pair of electrons, due to the exchange of an infinite number of virtual photons, including all possible arrangements of closed-fermion loops, is frequently expressed by the sum over an infinite number of Feynman graphs of increasing order and complexity. But the essential part of the same expression (before MSA) may be displayed in a simple, non-perturbative form by the functional expression

$$\mathrm{e}^{\mathcal{D}_A}G_{\text{c}}(x_1, y_1|A)G_{\text{c}}(x_2, y_2|A)\frac{\mathrm{e}^{L[A]}}{\langle S\rangle}\bigg|_{A\to 0} - (x_1 \leftrightarrow x_2),$$

from which one can readily understand the field-theoretic interest in obtaining quality approximations for $G_{\text{c}}[A]$ and $L[A]$.

2.6 Cluster decomposition

Consider a functional $L[A]$ which is acted upon by the linkage operator $\exp[\mathcal{D}]$, with, for this discussion, $\mathcal{D} = -\frac{\mathrm{i}}{2}\int \frac{\delta}{\delta A(u)}\Delta_{\text{c}}(u-v)\frac{\delta}{\delta A(v)}$. The notation is suggestive of QFT, but the discussion is quite general. For simplicity, we denote this operation upon $L[A]$ by $\bar{L}[A]$,

$$\bar{L}[A] = \mathrm{e}^{\mathcal{D}}L[A]. \tag{2.41}$$

The question posed here is how to represent the operation

$$S[A] = \mathrm{e}^{\mathcal{D}} \cdot \mathrm{e}^{L[A]}. \tag{2.42}$$

One method of approach is, using the techniques of the previous sections, to convert (2.42) to the FI

$$S[A] = \exp\left[-\frac{\mathrm{i}}{2}\int J\Delta_{\text{c}}J\right] \cdot N \int \mathrm{d}[\chi]$$

$$\cdot \exp\left[-\frac{\mathrm{i}}{2}\int \chi\Delta_{\text{c}}^{-1}\chi + \mathrm{i}\int \chi J + L[\chi]\right], \tag{2.43}$$

where $J = \int \Delta_{\text{c}}^{-1}A$. In appropriate situations, semi-classical methods may be designed to give an approximate evaluation of this FI.

There is another method of approach, with origins in the cluster expansions of Statistical Physics, which may sometimes be useful. This method converts

$S[A]$ directly to an exponential whose argument is an infinite sum over a set of "connected" quantities, in the form

$$S[A] = \exp\left[\sum_{N=1}^{\infty} Q_N/N!\right], \tag{2.44}$$

where the $Q_N[A]$ are connected functionals, defined as

$$Q_N[A] = e^{\mathcal{D}} L^N[A]|_{\text{conn}}$$

$$= \prod_{i>j=1}^{N} e^{\mathcal{D}_{ij}} \cdot \prod_{i=1}^{N} e^{\mathcal{D}_i} L[A_i]|_{\text{conn}, A_i \to A}$$

$$= \prod_{i>j=1}^{N} e^{\mathcal{D}_{ij}} \prod_{i=1}^{N} \bar{L}[A_i]|_{\text{conn}, A_i \to A}. \tag{2.45}$$

In obtaining (2.45), the trivial generalization of (2.9) to N factors of $L[A]$ has been written, using individual fields A_i (and followed by the final limit wherein all A_i are replaced by the same A); and where

$$\mathcal{D}_i = -\frac{i}{2} \int \frac{\delta}{\delta A_i} \Delta_c \frac{\delta}{\delta A_i}, \qquad \mathcal{D}_{ij} = -i \int \frac{\delta}{\delta A_i} \Delta_c \frac{\delta}{\delta A_j}.$$

The subscript "conn" indicates that at least one linkage must be retained between any and each pair of $L[A_k]$ terms; all terms without such linkages are to be discarded. For $N = 1$, (2.45) defines $Q_1[A] = \bar{L}[A]$.

A combinatoric derivation of this cluster expansion has been given in both HMF#1 and 2, and need not be repeated here. Rather, we shall use only a simpler scaling argument to produce a method of calculation for each and every Q_N; that argument has also appeared in HMF#2. One asks how such quantities would change if each $L[A]$ were multiplied by a constant factor λ; clearly, if $L \to \lambda L$, then $Q_N \to \lambda^n \cdot Q_N$, and, with (2.44), $S[A]$ becomes

$$\exp\left[\sum_{n=1}^{\infty} \frac{\lambda^n}{n!} Q_n\right] = e^{\mathcal{D}} \cdot e^{\lambda L},$$

or

$$\sum_{n=1}^{\infty} \frac{\lambda^n}{n!} Q_n = \ln[e^{\mathcal{D}} \cdot e^{\lambda L}]. \tag{2.46}$$

Each Q_N can now be obtained by calculating $(\partial/\partial\lambda)^N|_{\lambda} = 0$ on (2.46). For example,

$$Q_1 = [e^{\mathcal{D}} \cdot L e^{\lambda L}]/[e^{\mathcal{D}} \cdot e^{\lambda L}]|_{\lambda=D} = e^{\mathcal{D}} L = \bar{L}; \tag{2.47}$$

and

$$Q_2 = e^{\mathcal{D}} \cdot L^2 - (e^{\mathcal{D}} \cdot L)^2 = [e^{\mathcal{D}_{12}} - 1]\bar{L}[A_1]\bar{L}[A_2]|_{A_1=A_2=A}, \quad (2.48)$$

which gives expression to the "connected" subscript of (2.45). A pictorial representation of some of the higher Q_N and their weightings is given in HMF#2.

Notes

1 The origins and the development of a functional approach to QFT has benefited from the thought of many, including Symanzik, Fradkin, Zumino, and Sommerfield; but foremost among them, as the originator of the functional description for QFT, stands Julian Schwinger. Perhaps the best way of expressing the quality and quantity of his work is to point to the collection of 183 of his most relevant papers, from 1937 to 1976, compiled by M. Flato, C. Fronsdal, and K. A. Milton, as *Selected Papers of Julian Schwinger*, D. Reidel (1979); and to the subsequent collection by K. A. Milton, *A Legacy: Seminal Papers of Julian Schwinger*, World Scientific (2000), a grouping of some 43 additional papers. In these collections are gems of unchanging quality, including the sequence entitled "The Theory of Quantized Fields, I–V", and his 1951 masterpiece: "On Gauge Invariance and Vacuum Polarization".

  One must also point to the Quantum Action Principle invented by Schwinger, which generalizes all of those due to Fermat, Bernoulli, Euler, Lagrange, and Hamilton, who preceeded him. And, one might also remark that Schwinger was a kind and gentle man, one whose grasp and usage of the English language was mellifluous beyond compare. In one of the final lectures of his life, he gave a seminar at the University of Nottingham, entitled "The Greening of Quantum Field Theory: George and Me", in which he paid homage to George Green, and the functions he invented, and which Schwinger so modified to produce new descriptions for a variety of physical processes. For those readers who would enjoy an in-depth biography of Schwinger, see K. A. Milton, *Climbing the Mountain*, Oxford University Press (2000).

  In contrast to the "Lagrangian" approach of Schwinger, Feynman's path to functional methods can be characterized as "Hamiltonian", described in the text he coauthored with A. R. Hibbs, *Quantum Mechanics and Path Integrals*, McGraw-Hill (1965). The intuitive and special touch that was Feynman's can be seen in his early papers on QFT: R. P. Feynman, *Rev. Mod. Phys.* **20** (1948) 367; *Phys. Rev.* **76** (1949) 749 and 769; and in his book *Quantum Electrodynamics*, W. A. Benjamin (1962).

  The many and profound contributions of E. S. Fradkin, spread across five decades and a wide variety of subjects, deserve special mention. In this book a derivation (in Chapter 3) and repeated use are made of his GF representation, which, in effect, serves to move the formal operations of Schwinger into the realm of practical, functional calculations. The explanation of his techniques given in this book follows the notation used in HMF#2.

2 K. Symanzik, *Z. Naturforschung*, **92** (1954) 809.

3 B. Zumino, *NYU Lecture Notes* (1958).

4 C. Sommerfield, *Ann. Phys.* **26** (1963) 1.

5 B. Zumino, *Nucl. Phys.* B **89** (1975) 535.

6 F. A. Berezin, *Dolk. Acad. Nauk. SSSR* **137** (1961) 311.

3

Schwinger–Fradkin methods

The general construction of the functional solution to an inhomogeneous Green's function equation, such as that for a scalar function with scalar interaction, as defined in a 4-dimensional context,

$$[m^2 - \partial_x^2 + gA(x)]G_c(x, y|A) = \delta^{(4)}(x - y), \tag{3.1}$$

will be given in this chapter, along with the construction of corresponding Green's functions in QED and QCD. These are the generalizations of the "free-particle" propagators noted in (1.30), each defined in the presence of an arbitrary, "external" or "background" source $A(z)$; in every causal theory there is a way to express physical quantities of interest in terms of such Green's functions. Very similar techniques will be relevant if one is interested in Galilean rather than Lorentz symmetry; or if one requires the Green's function for interactions of arbitrary tensorial character, in arbitrary numbers of space–time dimensions; examples in QED and QCD will be sketched below.

3.1 Proper-time representations of Schwinger and Fradkin

It will be convenient to rewrite (3.1) as

$$[m^2 - \partial^2 + gA]G_c[A] = 1, \tag{3.2}$$

in which the formal operators ∂^2 and A are local and satisfy $\langle x|\partial^2|y\rangle = \partial_x^2 \delta(x-y)$, $\langle x|A|y\rangle = A(x) \cdot \delta(x - y)$, while the Green's function is given by the projection: $G_c(x, y|A) = \langle x|G_c[A]|y\rangle$. As mentioned in the first chapter, there are really four possible and independent solutions to (3.1), specified by the symbols c, $\bar{c}$, R, and A, depending on whether the desired solution is causal, anticausal, retarded, or advanced. For definiteness, simplicity, and usefulness, we shall consider only causal Green's functions, defined (as in every order of perturbation theory) by the replacement: $m^2 \rightarrow m^2 - i\epsilon, \epsilon \rightarrow 0+$.

33

It was Schwinger[1] who rewrote (3.2) in the form

$$G_c[A] = [m^2 - \partial^2 + gA - i\epsilon]^{-1}, \tag{3.3}$$

and introduced the exponential representation appropriate to a causal propagator, in a manner analogous to (1.27),

$$G_c[A] = i \int_0^\infty ds \, e^{-ism^2} \cdot e^{-is[gA-\partial^2]}. \tag{3.4}$$

The denominator of (3.3) is an hermitian operator with real eigenvalues, and hence, when diagonalized, each such term is reduced to the real-number analysis of (1.27). The parameter "s" over which integration is performed in (3.4) is called the "proper time", although it rarely carries the true dimensions of its name; in relativistic field theory its most important characteristics are that it is gauge invariant and Lorentz invariant.

One should observe that any perturbative expansion of (3.3) or (3.4) in powers of gA is exactly equivalent to the expansion of the integral equation for $G_c(x, y|A)$ in powers of gA,

$$G_c(x, y|A) = \Delta_c(x - y) - g \int d^4z \Delta_c(x - z)A(z)G_c(z, y|A),$$

or

$$G_c(x, y|A) = \Delta_c(x - y) - g \int d^4z G_c(x, z|A)A(z)\Delta_c(z - y),$$

where $G_c(x, y|0) = \Delta_c(x - y; m^2)$ denotes the free-particle propagator of (1.42). The intention of all that follows is to provide a representation of $G_c[A]$ which is exact, and which, if necessary, can be easily and reasonably approximated in a non-perturbative way.

The utility of such forms was demonstrated by Schwinger in his seminal paper,[2] where explicit solutions were produced for fermion GFs $G_c[A]$ and the associated, closed-fermion-loop functional $L[A]$ in the QED case of particle propagation in either a constant electromagnetic field or a laser/plane-wave field of a single frequency. A decade later, Fradkin[3] published an exact functional GF representation for particle propagation in an arbitrary, external field, one which can be generalized to all causal problems of any symmetry, and used to construct a variety of non-perturbative approximations to $G_c[A]$ and $L[A]$.

The derivation of Fradkin's representation begins by rewriting (3.4) in the apparently more-complicated form

$$G_c[A] = i \int_0^\infty ds \, e^{-ism^2} \left(e^{-i\int_0^{s+\delta} ds'[gA - \partial^2 + iv_\mu(s')\partial_\mu]} \right)_+ \Bigg|_{v_\nu \to 0, \, \delta \to 0}, \tag{3.5}$$

where $v_\mu(s')$ denotes an arbitrary vector function of the parameter s', and $(\ )_+$ is an OE, ordered with respect to s'; when v_μ vanishes in (3.5), the resulting OE is exactly equivalent to the ordinary exponential of (3.4).

The OE of (3.5), hereinafter called $U(s)$, displays the properties

$$\frac{\delta U}{\delta v_\mu(s)} = \partial_\mu U, \tag{3.6a}$$

and

$$\frac{\partial U}{\partial s} = -i[gA - \partial^2 + iv_\mu(s)\partial_\mu]U = -i\left[gA - \frac{\delta^2}{\delta v^2(s)} + iv(s)\cdot\partial\right]U. \tag{3.6b}$$

From its definition in (3.5), one sees that $U(0) = 1$. A solution satisfying these requirements is given by

$$U(s) = e^{i\int_0^s ds'\frac{\delta^2}{\delta v^2(s')}} \cdot \left(e^{\int_0^{s+\delta} ds'[v(s')\cdot\partial - igA]}\right)_+, \tag{3.7}$$

in the limit of $\delta \to 0+$, a limit that shall subsequently be understood; this limit is needed only to insure (in this formulation) the necessary commutation of $\exp[i\int_0^s ds'\frac{\delta^2}{\delta v^2(s')}]$ with $(v(s)\cdot\partial)$.

As discussed in the previous chapter, the exponential of a quadratic functional derivative, as in (3.7), is equivalent to an FI with Gaussian weight over the $v(s)$-dependent OE $V(s)$,

$$V(s) = \left(\exp\int_0^s ds'[v(s')\cdot\partial - igA]\right)_+, \tag{3.8}$$

and may easily be rewritten in the form

$$U(s) = N(s)\int d[\phi]\, e^{\frac{i}{4}\int_0^s ds'\phi_\mu^2(s')} \cdot V(s|\phi), \tag{3.9}$$

where $N(s)$ is the appropriate, Gaussian normalization factor,

$$N^{-1} = \int d[\phi]\exp\left[\frac{i}{4}\int_0^s ds'\phi_\mu^2(s')\right].$$

The OE $V(s)$ can be obtained explicitly, and we here sketch that construction by first writing

$$V(s|v) = e^{\int_0^s ds'v\cdot\partial} \cdot e^{\int_0^s v\cdot\partial}V(s|v) \equiv e^{\int_0^s ds'v\cdot\partial}\mathcal{F}(s), \tag{3.10}$$

so that $\mathcal{F}(s)$ satisfies the DE

$$\frac{\partial\mathcal{F}}{\partial s} = -ig\, e^{-\int_0^s ds'v\cdot\partial}Ae^{+\int_0^s ds'v\cdot\partial}\mathcal{F}(s). \tag{3.11}$$

It is now useful to take partial matrix elements of (3.11), applying $\langle x|$ to both sides of that equation, and using the properties $\langle x|\partial_\mu = (\partial/x_\mu)\langle x|$, and $\langle x|A = A(x)\langle x|$. One readily obtains

$$\frac{\partial}{\partial s}\langle x|\mathcal{F}(s) = -ig\, e^{-\int_0^s ds'v\cdot\partial^x} A(x)e^{+\int_0^s ds'v\cdot\partial^x}\langle x|\mathcal{F}(s)$$

$$= -ig\, e^{-\int_0^s ds'v\cdot\partial^x} A(x)\left(x + \int_0^s ds'v(s')\right)\mathcal{F}(s)$$

$$= -ig A\left(x - \int_0^s ds'v\right)\langle x|\mathcal{F}(s), \tag{3.12}$$

which is an explicit DE for $\langle x|\mathcal{F}(s)$, with solution

$$\langle x|\mathcal{F}(s)|y\rangle = e^{-ig\int_0^s ds'A(x-\int_0^{s'} v)}\delta^{(4)}(x - y), \tag{3.13}$$

after projecting both sides of this solution onto $|y\rangle$. From (3.10), matrix elements of $V(s)$ are immediately given, so that the complete GF may be written as

$$G_c(x, y|A) = i\int_0^\infty ds\, e^{-ism^2} e^{i\int_0^s ds'\frac{\delta^2}{\delta v^2(s')}} \cdot \delta\left(x - y + \int_0^s v\right)e^{-ig\int_0^s ds'A(y-\int_0^{s'} v)}. \tag{3.14}$$

For more complicated interactions, where, e.g., $A(z)$ is a matrix, it should be emphasized that the solution of (3.12) is an OE:

$$\langle x|\mathcal{F}|s)|y\rangle = \left(e^{-ig\int_0^s ds'A(x-\int_0^{s'} v)}\right)_+\langle x|y\rangle, \tag{3.15}$$

where the ordering is with respect to s'. Only when and if all quantities $A(z)$ bearing different s' values commute will the OE of (3.15) reduce to the ordinary exponential of (3.13).

Equation (3.14), or its equivalent FI obtained from (3.9), is the Fradkin representation for $G_c[A]$, appearing in an explicit form for a specified $A(z)$. All the complexity of the problem, due to the original lack of commutivity of ∂^2 and $A(x)$ in (3.3), has been replaced by Gaussian fluctuations over the v-dependence of (3.14), whose evaluation is limited only by the nature of the argument of A. It is immediately clear that any function $A(z)$ linear or quadratic in its argument will allow the functional operation of (3.14) to be carried through exactly. With few exceptions, a more complicated z-dependence of $A(z)$ will lead to a $G_c[A]$ that cannot be expressed exactly in closed form – that is, in terms of a finite number of quadratures – and the practical question then becomes one of finding simple, sensible, and (if at all possible) physically-intuitive ways of approximating this exact Fradkin representation.

3.2 Fradkin representations for QED and QCD

The basic DE for the GF of QED is

$$\left(m + \gamma_\mu\left[\partial_\mu^x - ig A_\mu(x)\right]\right) G_c(x, y|A) = \delta^{(4)}(x - y), \qquad (3.16)$$

which can be written in a symbolic, operator form (for any number of dimensions), as

$$G_c[A] = [m - i\gamma \cdot \Pi]^{-1}, \qquad (3.17)$$

where $\Pi_\mu = i(\partial_\mu - ig A_\mu)$. We shall use g to denote the (unrenormalized) electric charge. When $g \to 0$, $G_c[A] \to S_c$, the free-fermion propagator satisfying

$$(m + \gamma \cdot \partial) S_c(x - y) = \delta^{(4)}(x - y),$$

and given by

$$S_c(x - y) = (m - \gamma \cdot \partial^x)\Delta_c(x - y; m^2).$$

One can then write an integral equation for $G_c[A]$ in the form

$$G_c = S_c + ig S_c \gamma \cdot A G_c, \qquad (3.18)$$

or

$$G_c = S_c + ig G_c(\gamma \cdot A) S_c, \qquad (3.19)$$

using the convenient, formal notation. Equations (3.18) and (3.19) are equivalent in so far as their pertubative expansions are concerned; but for strong-coupling problems these forms are not particularly useful. The Dirac matrices here satisfy $\{\gamma_\mu, \gamma_\nu\} = 2\delta_{\mu\nu}$, and (no sum) $\gamma_\mu^2 = 1$.

Following Schwinger, one first rationalizes the denominator of (3.17) by rewriting it as

$$G_c = (m + i\gamma \cdot \Pi) \cdot [(m - i\gamma \cdot \Pi)(m + i\gamma \cdot \Pi)]^{-1},$$

or

$$G_c = (m + i\gamma \cdot \Pi) \cdot [m^2 + (\gamma \cdot \Pi)^2]^{-1}, \qquad (3.20)$$

where $(\gamma \cdot \Pi)^2 = \Pi^2 + ig\sigma_{\mu\nu} F_{\mu\nu}$, with $\sigma_{\mu\nu} = (\frac{1}{4})[\gamma_\mu, \gamma_\nu]$. Remembering that m is to have an infinitesimal, negative, imaginary part, as appropriate to the definition of this causal propagator, one introduces the representation

$$[m^2 + (\gamma \cdot \Pi)^2]^{-1} = i \int_0^\infty ds \, \exp\{-is[m^2 + (\gamma \cdot \Pi)^2]\}, \qquad (3.21)$$

where the "proper-time" variable really has the dimensions of (time)^2; frequently, the continuation $s \to -i\tau$ is made, and τ is then referred to as the proper time.

Before discussing Fradkin's representation of $\exp[-is(\gamma \cdot \Pi)^2]$, it will be useful to derive Schwinger's initial representation for the fermion closed-loop functional $L[A] = \text{Tr} \ln[1 - ig(\gamma \cdot A)S_c]$, which produces a form analogous to that of (3.21). Using the parametric representation of (2.16),

$$L[A] = -i \int_0^g dg' \, \text{Tr}\{(\gamma \cdot A)S_c[1 - ig'(\gamma \cdot A)S_c]^{-1}\},$$

or

$$L[A] = -i \int_0^g dg' \, \text{Tr}\{(\gamma \cdot A)G_c[g'A]\}, \tag{3.22}$$

one substitutes into (3.22) the form (3.20) and the representation (3.21), and discards all terms proportional to the vanishing (Dirac) trace over an odd number of γs to obtain

$$L[A] = i \int_0^\infty ds \, e^{-ism^2} \cdot \int_0^g dg' \, \text{Tr}\{(\gamma \cdot A)(\gamma \cdot \Pi)e^{-is(\gamma \cdot \Pi)^2}\}, \tag{3.23}$$

where the coupling constant inside Π is g'. Because of the trace operation, (3.23) can be rewritten as

$$L[A] = -\frac{1}{2} \int_0^\infty \frac{ds}{s} e^{-ism^2} \int_0^g dg' \frac{\partial}{\partial g'} \, \text{Tr}\{\exp[-is(\gamma \cdot \Pi)^2]\},$$

or

$$L[A] = -\frac{1}{2} \int_0^\infty \frac{ds}{s} e^{-ism^2} \text{Tr}\{\exp[-is(\gamma \cdot \Pi)^2]\} - (g \to 0), \tag{3.24}$$

where the coupling constant of Π is again g.

For both $G_c[A]$ and $L[A]$, the essential quantity to be understood is $U(s) = \exp[-is(\gamma \cdot \Pi)^2]$. A perturbative development, along with a solution for the two special cases of constant $F_{\mu\nu}$ and fields depending on a single frequency, was given by Schwinger;[2] but it is possible to arrange a non-perturbative approach with the aid of the representation introduced by Fradkin. For this, one replaces $U(s)$ by the seemingly more-complicated OE

$$U(s, v) = \left(\exp \left\{ -i \int_0^s ds'[\Pi^2 + ig\sigma \cdot F + v_\mu(s')\Pi_\mu] \right\} \right)_+, \tag{3.25}$$

with the property that $U(s, v)|_{v=0} = U(s)$. One notes that $U(s, v)$ satisfies the relations

$$\frac{\partial U}{\partial s} = -i[\Pi^2 + ig\sigma \cdot F + v(s) \cdot \Pi]U \tag{3.26}$$

and

$$\frac{\delta U}{\delta v_\mu(s)} = -i\Pi_\mu U \qquad (3.27)$$

and the initial condition $U|_{s=0} = 1$.

As in the scalar case, the reason for doing this is that $U(s, v)$ can be given a particularly elegant representation, of the form

$$U = \exp\left[i\int_0^{s-\delta} ds' \frac{\delta^2}{\delta v_\mu^2(s')}\right] W(s, v)|_{\delta \to 0},$$

with

$$W(s, v) = \left(\exp\left\{-i\int_0^s ds'[v_\mu(s')\Pi_\mu + ig\sigma \cdot F]\right\}\right)_+. \qquad (3.28)$$

And, as in the previous section, it is easy to see that (3.28) provides a solution to (3.26) and (3.27) in the limit $\delta \to 0$; and we shall assume this limit in everything that follows.

We now calculate, explicitly, configuration-space matrix elements of W, and begin by writing the LHS projection of W as

$$\langle x|W(s) = \langle x|\exp\left[\int_0^s ds' v_\mu(s')\partial_\mu\right]\mathcal{F}(s) = \exp\left[\int_0^s ds' v \cdot \partial^x\right]\langle x|\mathcal{F}(s).$$

$$(3.29)$$

Using the DE satisfied by $W(s)$, which may be read off from its definition in (3.28), one finds

$$\frac{\partial}{\partial s}\langle x|\mathcal{F}(s)$$
$$= -ig\, e^{-\int_0^s v \cdot \partial^x}[v(s) \cdot A(x) + i\sigma \cdot F(x)] \cdot e^{+\int_0^s v \cdot \partial^x}\langle x|\mathcal{F}(s)$$
$$= -ig\left[v(s) \cdot A\left(x - \int_0^s ds' v(s')\right) + i\sigma \cdot F\left(x - \int_0^s ds' v(s')\right)\right]\langle x|\mathcal{F}(s),$$

$$(3.30)$$

which is an explicit equation for $\langle x|\mathcal{F}(s)$ in terms of the fields, and has as its solution

$$\langle x|\mathcal{F}(s) = \left(\exp\left[-ig\int_0^s ds'\left\{v(s') \cdot A\left(x - \int_0^{s'} ds'' v(s'')\right)\right.\right.\right.$$
$$\left.\left.\left. + i\sigma \cdot F\left(x - \int_0^{s'} ds'' v\right)\right\}\right]\right)_+ \langle x|,$$

so that

$$\langle x|W(s)|y\rangle = \left(\exp\left[-ig\int_0^s ds'\left\{v(s')\cdot A\left(y - \int_0^{s'} v\right)\right.\right.\right.$$

$$\left.\left.\left. + i\sigma\cdot F\left(y - \int_0^{s'} v\right)\right\}\right]\right)_+ \cdot \delta\left(x - y + \int_0^s v\right). \quad (3.31)$$

Note that the OE is necessary in obtaining (3.28) only because of the s'-dependent (Dirac) matrix $\sigma\cdot F$. In two dimensions, however, there is just one antisymmetric matrix σ_{14} which can enter, while for a constant $F_{\mu\nu}$, $\sigma\cdot F$ becomes a constant (independent of s') matrix which commutes with itself and unity; and in both of these cases, the OE becomes an ordinary exponential.

With the aid of (3.20), (3.21), and the Fradkin solution of (3.28), one can write, finally, the representations

$$G_c(x, y|A) = i\int_0^\infty ds\, e^{-ism^2}\left(m - \gamma_\mu\frac{\delta}{\delta v_\mu(s)}\right)$$

$$\cdot \exp\left[i\int_0^s ds'\frac{\delta^2}{\delta v^2(s')}\right]\cdot \langle x|W(s)|y\rangle|_{v\to 0} \quad (3.32)$$

and

$$L[A] = -\frac{1}{2}\int_0^\infty \frac{ds}{s}\, e^{-ism^2} \mathrm{tr}\cdot\int d^D x\cdot e^{i\int_0^s ds'\frac{\delta^2}{\delta v^2(s')}}$$

$$\cdot \{\langle x|W(s)|x\rangle - \langle x|W(s)|x\rangle|_{g=0}\}|_{v\to 0}, \quad (3.33)$$

where tr denotes a trace over Dirac coordinates, and in which all the field dependence is explicit.

3.3 Gauge structure in QED and QCD

In order to isolate the gauge-variant part of $\langle x|W|y\rangle$ – that is, those parts which change under the $U(1)$ gauge transformations of QED, $A_\mu \to A_\mu + \partial_\mu\Lambda$ – it is useful to consider the quantity

$$Q(\lambda) \equiv \int_0^s ds'v_\mu(s')A_\mu\left(y - \lambda\int_0^{s'} v\right),$$

which for $\lambda = 1$ appears in the exponential of (3.31). A simple integration-by-parts produces

$$Q(\lambda) = A_\mu\left(y - \lambda\int_0^s v\right)\cdot\int_0^s ds'v_\mu(s')$$

$$+ \lambda\int_0^s ds'\int_0^{s'} ds''v_\mu(s'')v_\nu(s')\partial_\nu A_\mu\left(y - \lambda\int_0^{s'} v\right), \quad (3.34)$$

but, because of the $\delta(x - y + \int_0^s v)$ factor of (3.31), the first RHS term of (3.34) may be replaced by $(y - x)_\mu A_\mu(\lambda x + [1 - \lambda]y)$. Then, with the definition of $F_{\mu\nu}$, (3.34) can be rewritten as

$$Q(\lambda) = (y - x)_\mu A_\mu(\lambda x + (1 - \lambda)y)$$

$$+ \lambda \int_0^s ds' \int_0^{s'} ds'' v_\mu(s'') v_\nu(s') \left[F_{\nu\mu} \left(y - \lambda \int_0^{s'} v \right) \right.$$

$$\left. + \partial_\mu A_\nu \left(y - \lambda \int_0^{s'} v \right) \right]. \tag{3.35}$$

But the last RHS term of (3.35) may be replaced by

$$-\lambda \int_0^s ds' v_\nu(s') \frac{\partial}{\partial \lambda} A_\nu \left(y - \lambda \int_0^{s'} v \right) = -\lambda \frac{\partial}{\partial \lambda} Q(\lambda),$$

which, when substituted into (3.35), produces the DE

$$\frac{\partial}{\partial \lambda}(\lambda Q(\lambda)) = (y - x)_\mu A_\mu(\lambda x + (1 - \lambda)y)$$

$$+ \lambda \int_0^s ds' \int_0^{s'} ds'' v_\mu(s') v_\nu(s'') F_{\nu\mu} \left(y - \lambda \int_0^{s'} v \right). \tag{3.36}$$

The integral of (3.36) between $\lambda = 0$ and $\lambda = 1$ is immediate, and yields

$$Q(\lambda) = - \int_y^x d\xi_\mu A_\mu(\xi)$$

$$- \int_0^1 \lambda \, d\lambda \cdot \int_0^s ds' \int_0^{s'} ds'' v_\mu(s') v_\nu(s'') F_{\mu\nu} \left(y - \lambda \int_0^{s'} v \right), \tag{3.37}$$

where ξ_μ denotes the straight-line path between x_μ and y_μ, $\xi_\mu = \lambda x_\mu + (1 - \lambda)y_\mu$.

By this simple computation, one sees that $L[A]$ is gauge invariant; and that the only gauge-variant part of $G_c(x, y|A)$ is that factor coming from (3.37), $\exp[ig \int d\xi_\mu A_\mu(\xi)]$. Under the gauge change $A_\mu \to A'_\mu = A_\mu + \partial_\mu \Lambda$, the only variation of the complete $G_c[A]$ is

$$G_c(x, y|A + \partial \Lambda) = e^{ig[\Lambda(x) - \Lambda(y)]} G_c(x, y|A), \tag{3.38}$$

a property which may be inferred directly from the equations which define $G_c[A]$.

It will also be useful to comment on the corresponding gauge properties found in QCD, with the definitions of $G_c[A]$ and $L[A]$ the same as in QED except for the replacement $A_\mu \to A_\mu^a \lambda^a$, where the λ^a are the fundamental or defining matrix representations of $SU(N)$ (for $N = 3$, the Gell–Mann matrices) satisfying

$$[\lambda^a, \lambda^b] = 2\mathrm{i} f_{abc}\lambda^c, \qquad \{\lambda^a, \lambda^b\} = \frac{4}{N}\delta_{ab} + 2d_{abc}\lambda^c,$$

$$\mathrm{tr}[\lambda^a] = 0, \qquad \mathrm{tr}[\lambda^a\lambda^b] = 2\delta_{ab}. \tag{3.39}$$

Following from an original Lagrangian density of the form

$$\mathcal{L} = -\frac{1}{4}\left(F_{\mu\nu}^a\right)^2 - \bar{\psi}\left[m + \gamma_\mu\left(\partial_\mu - \mathrm{i}gA_\mu^a\lambda^a\right)\right]\psi,$$

local, position-dependent gauge transformations which leave this quark–gluon Lagrangian invariant are given by

$$A_\mu^a(z)\lambda^a \equiv A_\mu(z) \to A_\mu'(z) = V^+(z)\left(A_\mu(z) + \frac{\mathrm{i}}{g}\partial_\mu\right)V(z),$$

$$F_{\mu\nu}^a(z)\lambda^a \equiv F_{\mu\nu}(z) \to F_{\mu\nu}'(z) = V^+(z) \cdot F_{\mu\nu}(z) \cdot V(z), \tag{3.40}$$

for arbitrary $V(z) = \exp[\mathrm{i}\lambda^a\omega^a]$, where $F_{\mu\nu}^a = \partial_\mu A_\nu^a - \partial_\nu A_\mu^a + gf_{abc}A_\mu^b A_\nu^c$. The Fradkin representations for $G_c[A]$ and $L[A]$ go through as before, except that the color matrix factors λ^a require the use of OEs everywhere in the formula corresponding to (3.31), and the tr operation includes a summation over such color variables.

The invariance of $L[A]$ under the full transformation (3.40) can be shown in a simple way, by writing $\langle x|W(s)|x\rangle = U(x)\delta(\int_0^s ds'v(s'))$ and seeing how $U(s)$ changes under a gauge transformation, where $U(s) \to U'(s)$, and

$$U'(s) = \left(\exp\left[-\mathrm{i}g\int_0^s ds'\left\{v_\mu(s')V^+\left(x - \int_0^{s'} v\right)\left(A_\mu + \frac{\mathrm{i}}{g}\partial_\mu\right)V\right.\right.$$

$$\left.\left. + \mathrm{i}\sigma_{\mu\nu}V^+\left(x - \int_0^{s'} v\right)F_{\mu\nu}V\right\}\right]\right)_+, \tag{3.41}$$

and where, in writing (3.41) and similar expressions, dependence on the common variable $x - \int_0^{s'} ds''v(s'')$ is exhibited only in the first term of any product. To understand the relation between U and U', it is useful to consider the DE for U',

$$\frac{\partial U'}{\partial s} = -\mathrm{i}g\left[v_\mu(s)V^+\left(x - \int_0^s v\right)\left(A_\mu + \frac{\mathrm{i}}{g}\partial_\mu\right)V\right.$$

$$\left. + \mathrm{i}\sigma_{\mu\nu}V^+\left(x - \int_0^s v\right)F_{\mu\nu}V\right] \cdot U'. \tag{3.42}$$

Setting $U' = V^+(s) \cdot Z(s)$, with $V(s) \equiv V(x - \int_0^s v)$, substitution into (3.42) generates the DE for Z,

$$\frac{\partial Z}{\partial s} = -ig\left[v_\mu(s)A_\mu\left(x - \int_0^s v\right) + i\sigma_{\mu\nu}F_{\mu\nu}\left(x - \int_0^s v\right)\right] \cdot Z, \quad (3.43)$$

where the replacements $v_\mu(s)V^+(s)\partial_\mu V(s) = -V^+(s)\frac{\partial V}{\partial s}$ and $\frac{\partial V^+}{\partial s}V = -V^+\frac{\partial V}{\partial s}$ have been made.

Taking into account the initial condition $Z(0) = V(0)$, and in comparison with the equation and solution for $U(s)$, one can write the solution of (3.43) as $Z(s) = U(s)V(0)$, so that, finally, $U'(s) = V^+(s)U(s)V(0)$. But $V^+(s) = V^+(0)$, by the closed-loop condition $\int_0^s ds' v(s') = 0$, as required by the representation (3.33). Hence tr$[U'] =$ tr$[U]$, and $L[A]$ has been shown to be invariant under the full gauge transformations of QCD. This is not a surprise, of course, for the Fradkin representation is exact; but it is useful to see how the exact gauge property is fulfilled before attempting any approximations.[4]

3.4 Soluble examples: quadratic forms and perturbative approximations

Consider first the scalar-interaction $G_c[A]$ of (3.14). Because the most complicated functional integral that can be performed exactly is Gaussian – as is the case for ordinary integration – and because, as shown in Chapter 2, the linkage operation defining the Fradkin representation is equivalent to Gaussian-weighted functional integration, there follows from (3.14) the immediate rule-of-thumb for solubility: any $A(x)$ which is constant, linear, or quadratic in x leads immediately to an explicitly soluble $G_c(x, y|A)$ in the sense that the functional operations may be carried through in closed form. The result may still require the solution of an integral equation, but that is a different matter; "solubility" here means that the functional operations may be carried through without approximation.

For the case of QED, the $G_c(x, y|A)$ of (3.32) displays similar features: any $A_\mu(x)$ which is linear in x_ν is, in this sense, exactly soluble. This means that a constant $F_{\mu\nu}$ generates a soluble result, as first noted and calculated by Schwinger[2] many years ago; and we shall reproduce his result in the next section, along with a non-trivial generalization. Almost identical remarks may be made for the $G_c(x, y|A)$ of QCD, except that here an $A_\mu^a(x)$ linear in x_ν does not imply a constant $F_{\mu\nu}^a$. This means that the Fradkin linkage operation must be carried out upon Gaussian v_μ dependence inside the OE, a procedure which appears formidable, but which can be carried through at least partially, as in the context of the quasi-Abelian limit of Chapter 8.

Fortunately, experience has taught that there are other exactly soluble problems, which have nothing to do with the Gaussian rule-of-thumb. These solutions are associated with potentials $A(x)$ that either represent plane-wave (or "laser") fields, or have a restriction on the type of x-dependence they carry, viz.: $A(x) \to A(x_3 - x_0)$. Such models can be used to represent and calculate sums over significant subsets of Feynman graphs for high-energy (eikonal) scattering problems, while their generalizations can point the way to the development of new and useful approximation methods. For example, the "scalar laser" solution of Section 4.2, for which full solubility of $G_c(x, y|A)$ is obtained if $A(x) \to A(k \cdot x)$, when $k^2 = \sum_{\mu=1}^{4} k_\mu^2 = 0$ can be looked upon as the zeroth term in an approximation for which $-k^2 = \mu^2 \neq 0$, but for which $(\mu/m) \ll 1$. And the forms one finds bear a strong resemblance to those appearing in the non-soluble situation wherein $A(x) \to A(x_3 + x_0, x_3 - x_0)$, a function of both light-cone variables.

Ordinary perturbation theory represents one systematic approximation method which is always "soluble", in the functional sense, because any expansion of these $G_c[A]$ in powers of the coupling constant leads to linkage operations which can be trivially performed by repeated use of

$$
e^{-\frac{1}{2}\int \frac{\delta}{\delta A} D_c \frac{\delta}{\delta A}} A(x_1) \cdots A(x_n)|_{A=0}
$$

$$
= -\mathrm{i} e^{-\frac{1}{2}\int \frac{\delta}{\delta A} D_c \frac{\delta}{\delta A}} \int D_c(x_1 - u) \frac{\delta}{\delta A(u)} A(x_2) \cdots A(x_n)\bigg|_{A \to 0}
$$

$$
= -\mathrm{i} \sum_{j=2}^{n} D_c(x_1 - x_j) e^{-\frac{1}{2}\int \frac{\delta}{\delta A} D_c \frac{\delta}{\delta A}} [A(x_2) \cdots A(x_n)]_{j, A=0},
$$

where $[A(x_2) \cdots A(x_n)]_j$ contains all $n - 2$ factors $A(x_i)$ except that A with $i = j$.

Perturbation theory applied to QED has been most useful in developing our trust in QFT as a fundamental expression of the quantum world – comparisons of experiment and theory in agreement to more than eight significant figures cannot be dismissed – but the perturbative approach should always be viewed with suspicion, especially in the light of Dyson's famous observation:[5] Vacuum structure in QED, expressed as some $f(e^2)$, cannot be expected to possess a convergent expansion within a circle, however small, about $e = 0$, for such a region involves negative values of e^2, for which an e^+ and an e^- would repel each other; here, the vacuum would have no lower energy bound, and would spontaneously tend to decay into charged pairs. In fact, for all the known examples, exact and approximate, of charged pairs torn from the vacuum by intense external fields, one finds expressions for the vacuum persistence probability (or equivalently, the probability of pair production) which contain an essential singularity in the coupling constant at $e = 0$, and to which perturbative techniques cannot be applied.

3.5 Pair production in generalized electric fields

One of the best-known, non-trivial examples of a QED process which can be carried out without approximation is Schwinger's 1951 calculation[2] of the "vacuum persistence probability" corresponding to the probability for the production of oppositely-charged pairs in the presence of a constant, external field. That calculation is rather more potential theory than field theory, in that it neglects radiative corrections of the photon field, replacing (2.45) by $\langle 0|S|0 \rangle = \exp\{L[A]\}$, where A denotes the external vector potential. Schwinger's result has since been obtained in a variety of ways,[6] along with some generalizations; from the present, functional point-of-view, one can begin from the $L[A]$ of (3.33) and (3.37). The gauge-invariant result is "soluble" because the Fradkin linkage operation is Gaussian; but different formulations and different gauges will display differing levels of complexity in arriving at the same result.

In this section we give a new treatment suggested by the non-trivial generalization recently discovered by Tomaras, Tsamis, and Woodard;[7] it is not only simpler than previous functional calculations but, when used for selected $G_c[A]$, can provide structures useful for qualitative descriptions of other processes, as in the eikonal scattering model of Section 8.4.

The essence of the model is the restriction of the argument of A_μ to the light-cone coordinate $x_{(+)} = x_3 + x_0 : A_\mu(x) \to A_\mu(x_{(+)})$, corresponding to fields that "propagate" in the $-x_3$ direction. For simplicity, we suppress any associated magnetic fields or electric fields in perpendicular directions, with the choice $A_{1,2} = 0$; and choose the gauge specified by $A^{(-)} = A_3 - A_0 = 0$. In terms of the "light-cone projectors", $n_\mu^{(\pm)} = (0, 0, 1; \mp i)$, with the properties $[n^{(\pm)}]^2 = 0$, $n^{(+)} \cdot n^{(-)} = 2$, these statements may be rephrased as: $n^{(+)} \cdot A = A^{(+)}(x_{(+)})$, $n^{(-)} \cdot A = 0$, and $x_{(+)} = n^{(+)} \cdot x$. The electric field $E(x^{(+)})$ in the z-direction is then given by $-(d/dx_{(+)})A^{(+)}(x_{(+)})$.

Using (3.31) and (3.33), we begin by writing the complete expression for $L[A]$,

$$L[A] = -\frac{1}{2} \int_0^\infty \frac{ds}{s} e^{-ism^2} \int d^4x \int \frac{d^4p}{(2\pi)^4} e^{i\int_0^s ds' \sum_\mu \frac{\delta^2}{\delta v_\mu(s')^2}} \cdot e^{i\int_0^s v \cdot p}$$

$$\cdot e^{-ig\int_0^s ds' v_\mu(s')A_\mu(x^{(+)} - \int_0^{s'} ds'' n^{(+)} \cdot v(s''))} \cdot (e^{g\int_0^s ds' \sigma \cdot F(x^{(+)} - n^{(+)} \cdot \int_0^{s'} v)})_+ - (g \to 0),$$

$$(3.44)$$

and then, as we shall employ in various contexts in subsequent chapters, introduce unity under these integrands in the form

$$1 = \int d[u]\delta\left[u(s') - n^{(+)}_\mu \int_0^{s'} ds'' v_\mu(s'')\right]$$

or

$$1 = N' \int d[u] \int d[\Omega] e^{i\int_0^s ds' u(s')\Omega(s')} \cdot e^{-i\int_0^s ds' \Omega(s')n^{(+)}\cdot\int_0^{s'} ds'' v(s'')}, \tag{3.45}$$

where N' is a normalization constant. Abel's trick can be used to rewrite the second exponential factor of the RHS of (3.45) as

$$-i\int_0^s ds' v(s') \cdot n^{(+)} \int_{s'}^s ds'' \Omega(s'').$$

All of the v-dependence inside the arguments of A_μ and $F_{\mu\nu}$ has now been "extracted", and replaced by appropriate factors of $u(s')$, so that the complete Fradkin linkage operation takes the form

$$e^{i\int_0^s ds' \frac{\delta^2}{\delta v^2}} \cdot e^{i\int_0^s ds' v_\mu(s')[p_\mu - gA_\mu(x^{(+)} - u(s')) - n^{(+)}_\mu \int_{s'}^s ds'' \Omega(s'')]}\Big|_{v\to 0},$$

which can be evaluated exactly as

$$\exp\left\{-i\int_0^s ds'\left[p - gA(x^{(+)} - u(s')) - n^{(+)}\int_{s'}^s \Omega\right]^2\right\},$$

or as

$$\exp\left\{-i\int_0^s ds'\left[p^2 + g^2 A^2 - 2gp\cdot A - 2p^{(+)}s'\Omega(s') + 2gA^{(+)}(s')\int_{s'}^s \Omega\right]\right\} \tag{3.46}$$

using the property that $[n^{(+)}]^2 = 0$, which conveniently removes all quadratic exponential dependence on the Ω. (Were this not the case, the functional integration over Ω could still be performed, but not the subsequent one over $u(s')$.) Because $A_{1,2} = 0$, $p \cdot A$ may be written as $p_3 A_3 - p_0 A_0 = \frac{1}{2}[p^{(-)}A^{(+)} + A^{(-)}p^{(+)}]$; and since we have chosen the gauge $A^{(-)} = 0$, $p \cdot A \to p^{(-)}A^{(+)}/2$. For the same reasons $A^2 = 0$; and using Abel's trick in reverse, functional integration over Ω takes the form

$$N' \int d[\Omega] e^{i\int_0^s ds' \Omega(s')[u(s') + 2s'p^{(+)} - 2g\int_0^{s'} ds'' A^{(+)}(x^{(+)} - u(s''))]}$$

$$= \delta\left[u(s') + 2s'p^{(+)} - 2g\int_0^{s'} ds'' A^{(+)}(x^{(+)} - u(s''))\right],$$

which means that the only $u(s')$ which can enter must be a solution of the nonlinear integral equation

$$u(s') + 2s'p^{(+)} = 2g \int_0^{s'} ds'' A^{(+)}\big(x^{(+)} - u(s'')\big). \qquad (3.47)$$

One can now evaluate $\int d[u]$ in terms of the solutions $u(s')$ to (3.47) by making a change of (functional) variable to a new $f(s')$, defined as the argument of the delta-functional of (3.47), where

$$\frac{\delta f(s_1)}{\delta u(s_2)} = \delta(s_1 - s_2) - 2g\theta(s_1 - s_2)E\big(x^{(+)} - u(s_2)\big)$$

and

$$\frac{\partial}{\partial x^{(+)}} A^{(+)}\big(x^{(+)} - \mathcal{Z}\big) = -E\big(x^{(+)} - \mathcal{Z}\big).$$

Hence, the Jacobian of the transformation from $\int d[u]$ to $\int d[f]$ is given by

$$\det \left| \frac{\delta u}{\delta f} \right| = \exp[-\operatorname{Tr}\ln(1 - 2g \cdot \theta \cdot E)], \qquad (3.48)$$

where $\langle s_1 | \theta \cdot E | s_2 \rangle \equiv \theta(s_1 - s_2)E(x^{(+)} - u(s_2))$, and the $\int d[u]\delta[f]\mathcal{F}[u]$ can be written as

$$e^{-\operatorname{Tr}\ln(1 - 2g\theta E)}\mathcal{F}[u], \qquad (3.49)$$

where the $u(s')$ of (3.48) are restricted to the solutions of (3.47). Here, $\mathcal{F}[u]$ refers to all u-dependence arising from (3.46) and from the OE. Finally, there is a lovely simplification arising from the "retarded" nature of the θ-function of (3.38): only the first term in an expansion of the Trace log in powers of g can be non-zero; and hence, with $\theta(0) = 1/2$, the RHS of (3.48) may be replaced by

$$e^{+g \int_0^s ds' E\big(x^{(+)} - u(s')\big)}. \qquad (3.50)$$

Setting $s' = s$ in (3.47), one sees that the needed exponential factor $igp^{(-)} \int_0^s ds' A^{(+)}(x^{(+)} - u(s'))$ of (3.46) may be replaced by $ip^{(-)}[\frac{u(s)}{2} + sp^{(+)}]$. Further, for a purely electric field, the trace of the OE collapses into

$$4\cosh\left[g \int_0^s ds' E\big(x^{(+)} - u(s')\big)\right],$$

as is easily seen by summing all the non-zero terms of its expansion in powers of g. Thus, the functional parts of the calculation are completely specified by the solutions $u(s')$ to (3.47); and there remain only the subsequent integrations over x, p, and s.

There are now two distinct ways of proceeding: (i) first find the solutions $u(s')$ and then calculate $\int d^4 p$; or (ii) the converse. For a constant E, the first

path is the simplest, and both of its operations are trivial. However, path (ii) is
much more appropriate, since it immediately gives the answer for any varying
$E(x_{(+)})$; and it is this route which we here follow, by first performing

$$\int d^4 p \, e^{-isp^2 + igp^{(-)} \int_0^s ds' A^{(+)}(x^{(+)} - u(s'))} \mathcal{F}[u], \qquad (3.51)$$

where $u(s') = u(s'|p^{(+)})$ is the solution of (3.47), and

$$\mathcal{F}[u] = e^{g \int_0^s ds' E(x^{(+)} - u(s'))} \cdot \cosh\left[g \int_0^s ds' E\left(x^{(+)} - u(s')\right)\right].$$

Now (3.51) can be rewritten in the form

$$\int d^2 p_\perp e^{-isp_\perp^2} \cdot \frac{1}{2} \int dp^{(+)} \int dp^{(-)} e^{-isp^{(+)} p^{(-)}} \cdot e^{ip^{(-)}\left[\frac{1}{2} u(s|p^{(+)}) + sp^{(+)}\right]} \cdot \mathcal{F}[u[p^{(+)}]]$$

$$= \left(-i\frac{\pi}{s}\right) \cdot \frac{1}{2} \int dp^{(+)} \mathcal{F}[u] \int dp^{(-)} e^{ip^{(-)} u(s|p^{(+)})/2}$$

$$= -\frac{2i\pi^2}{s} \int dp^{(+)} \mathcal{F}[u[p^{(+)}]] \delta\left(u(s|p^{(+)})\right), \qquad (3.52)$$

and instead of attempting to integrate directly over $p^{(+)}$ it is convenient to make
a change of variable to $u(s|p^{(+)})$, defined implicitly in terms of $p^{(+)}$ by

$$u(s|p) = -2sp + 2g \int_0^s ds' A^{(+)}\left(x^{(+)} - u(s'|p)\right), \qquad (3.53)$$

where we have dropped the superscript of $p^{(+)}$. With the fixed "initial condition"
$u(0|p) = 0$, obvious from (3.53), the latter may be replaced by a first-order differential equation (DE) whose solution, for a specified $A^{(+)}(x)$, is unambiguous.
Upon changing variables from p to $u(s|p)$, one requires the integral equation
constructed by variation of (3.53) with respect to p,

$$J(s|p) = -2s + 2g \int_0^s ds' E\left(x^{(+)} - u(s'|p)\right) J(s'|p), \qquad (3.54)$$

where $J(s'|p) = (d/dp)u(s'|p)$. The DE corresponding to the s-variations of
(3.54) is

$$J'(s|p) = -2 + 2g E\left(x^{(+)} - u(s|p)\right) \cdot J(s|p), \qquad (3.55)$$

which, with the initial condition $J(0|p) = 0$, is equivalent to (3.54).

But (3.52) requires that for every p, $u(s|p) = 0$, so that (3.55) simplifies to

$$J'(s|p) = -2 + 2g E\left(x^{(+)}\right) \cdot J(s|p),$$

which has the immediate solution

$$J(s) = -2e^{gsE(x^{(+)})} \cdot \frac{\sinh\left(gsE(x^{(+)})\right)}{gsE(x^{(+)})},$$

and is independent of p. Hence (3.52) can be written as

$$-\frac{i\Pi^2}{s} \cdot J^{-1}(s)\mathcal{F}[u(s'|p)]|_{u(s|p)=0}, \qquad (3.56)$$

where the value of p to be used in (3.56) is obtained from (3.53) by setting $u(s|p) = 0$, namely

$$p = \frac{g}{s} \int_0^s ds' A^{(+)}(x^{(+)} - u(s'|p)). \qquad (3.57)$$

In fact, the situation is somewhat simpler than this. Had we considered path (i), for example for a constant E, we would have found a specific $u(s|p) = \lim u(s'|p)|s' \to s$ which is non-zero, as is the $u(s'|p)$, $s' < s$. And, as stated above, the $\int d^4p$ is then immediate, leading to Schwinger's result. But once the requirement is made, by first integrating over p^-, that $u(s|p) = 0$, then it follows from (3.53) that $u(s'|p) = 0$ for all s' (although J, the variation of u with respect to p, is non-zero). This can be seen by calculating $(d/ds')u(s'|p)$, which generates the relation

$$u'(s') + 2p = 2gA^{(+)}(x^{(+)} - u(s')), \qquad (3.58)$$

which we use to fix $u'(s')$ – suppressing the p-dependence for the moment – at different values of s' between 0 and s. For example, if $u(0) = u(s) = 0$, then from (3.58) the slopes $u'(0) = u'(s)$. Assume the slopes are positive (or negative). There must then be one point in the interval 0 to s, call it s_1, where the function $u(s_1)$ vanishes with slope opposite to that of $u'(0)$ and $u'(s)$. But if there is such a vanishing point, $u(s_1) = 0$, then from (3.58) one sees that it has the same slope as at 0 and s; and therefore each region, 0 to s_1 and s_1 to s, must contain a vanishing $u(s_2)$ of opposite slope; etc. But all such points of vanishing $u(s_k)$ must have the same slope, etc.; and hence the only possibility is that $u(s') = 0$. Restoring the p-dependence to $u(s')$, (3.57) simplifies to $p = A^{(+)}(x^{(+)})$, which is the value of $p^{(+)}$ obtained from integration over the delta function of $u(s|p)$.

Putting everything together, and with the aid of (3.56), one obtains

$$2L[A] = \frac{ig}{4\pi^2} \int d^4x E(x^{(+)}) \int_0^\infty \frac{ds}{s^2} e^{-ism^2} \frac{\cosh\left(gsE(x^{(+)})\right)}{\sinh\left(gsE(x^{(+)})\right)} - (g \to 0),$$

$$(3.59)$$

in which the electric field is now an arbitrary function of $x_{(+)}$. Note that the sign of $E(x_{(+)})$ is irrelevant, and for simplicity it will be taken as positive, while its dependence upon $x_{(+)}$ will be suppressed. Rotating contours such that, in effect, $s \to -i \cdot t$, with t real and positive and integrated between 0 and $+\infty$, one sees that singularities of the integrand arise at the points $t \to t_n - i\epsilon$, $t_n = n\pi/gE$; and with the aid of the familiar relation: $[t - (t_n - i\epsilon)]^{-1} = P/[t - t_n] - i\pi\delta(t - t_n)$, one easily finds

$$2\mathrm{Re}L[A] = -\frac{\alpha}{\pi^2} \int d^4x E^2 \sum_{n=1}^{\infty} \frac{1}{n^2} e^{-n\pi m^2/gE}, \tag{3.60}$$

which, with $\alpha = g^2/4\pi$, is Schwinger's 1951 result for the vacuum persistence probability, but here generalized to an electric field E which is an arbitrary function of $x_{(+)}$. (Note that the $n = 0$ term of the sum is missing, because of the necessary subtraction of any $g = 0$ dependence.)

The fact that this is an old result, at least in form, should not detract from its importance as an example of an exact calculation every term of which result is intrinsically non-perturbative about $g = 0$. Other, more recent calculations of analogous production processes[6] show a similar behavior, displaying essential singularities of various forms, as do instanton approximations to vacuum structure in a variety of field theories.[8] It should be noted that Dyson's observation was made concerning vacuum structure in the context of field-theoretic fluctuations of the quantized electromagnetic field, while the present singularities are associated with particle production from the vacuum due to an external, classical field. But both situations involve the vacuum, and lead to the clear conclusion that interactions associated with vacuum structure are intrinsically non-perturbative.

Notes

1 See Note 1, Chapter 2.
2 J. Schwinger, *Phys. Rev.* **82** (1951) 664.
3 E. S. Fradkin, *Nucl. Phys.* **76** (1966) 588.
4 The gauge invariance of IR approximations in QCD has been discussed in HMF#2, Chapter 12.
5 F. J. Dyson, *Phys. Rev.* **85** (1952) 631.
6 See, for example, M. N. Hounkonnou and M. Naciri, *J. Phys. G: Nucl. Part. Phys.* **26** (2000) 1849; and the many references given therein.
7 T. N. Tomaras, N. C. Tsamis, and R. P. Woodard, *Phys. Rev. D* **62** (2000)125005; hep-th/0007166. An improved derivation, and its application to $1 + 1$ dimensions, has been posted at hep-th/0108090.
8 Many references to instantons and large-order perturbation theory, in QED, QCD, QM and Critical Exponents, may be found in an unpublished HET Brown Report by P. F. Mende.

4

Lasers and crossed lasers

The title of this chapter is really a misnomer, for the word "laser" should properly be replaced by "electromagnetic plane wave" (epw); but we ask the reader's indulgence for this simple idealization, which is reasonable as long as the perpendicular dimensions of the laser beams under question, henceforth called the "width", are much larger than the dimensions of the charged particle on which they are acting, or than the transverse distances over which the particle is to move. The latter condition, in particular, is not always satisfied, and provides a limit of applicability of the idealization, as in the first section, below. Certainly, for the case of charged-pair production in the overlap region of volume D^3 of two perpendicularly-oriented laser beams, each of width D, one expects this idealization to be reasonable as long as $D > u_0 \lambda_c$, where λ_c denotes the Compton wavelength of each of the produced particles, and $u_0 \sim 10^2$ sets the scale for distances over which coherent absorption of the laser photons can take place.

4.1 Classical charged-particle propagation in a laser (epw) field

This problem can be and has been solved in several ways;[1] and its solution is worth discussing here, before quantum-mechanical treatments are begun, in order to provide a simple example of the utility of OEs, and to set the stage for what may occur after a pair is torn from the vacuum in the region of intersection of two intense lasers.

The Physics of the problem is simple, especially if the laser beam is imagined end-on, moving in the $+\hat{z}$ direction (into the page), with electric field E in the $+\hat{y}$ direction and in-phase magnetic field B pointing in the $+\hat{x}$ direction. As it acts on the charge, initially assumed at rest, the E field causes an upwards velocity $v_\perp$, and the B field then provides a force proportional to $v_\perp \times B$, in the

51

same direction as the beam, which continues until this part of the wave passes over the particle (which is moving with velocity $<c$) and the E and B fields at the position of the particle reverse. Because the particle has inertia, its $v_\perp$ can only be decreased gradually, while the reversal of B causes a deceleration of its forward motion, so that by the time that E and B are again zero, the particle's longitudinal motion has been stopped, and it finds itself a distance down the beam depending upon the intensity of the beam. As stated above, we must make the restriction that the particle's transverse velocity never takes it out of the beam; and we also neglect the (classical) radiation which must occur when the charge is accelerated. Other, practical limitations are discussed in the reference in Note 1; but it is clear that a sufficiently intense laser of sufficient width can provide a charged particle with enormous velocities over macroscopic distances.

We begin with the classical DE for a particle of charge e and mass m in an electromagnetic field, $F_{\mu\nu}(x, \tau)$,

$$\frac{d^2 x_\mu}{d\tau^2} = \frac{e}{m} \sum_{\nu=1}^{4} F_{\mu\nu}(x, \tau) \cdot \frac{dx_\nu}{d\tau}, \tag{4.1}$$

where τ is the particle's proper time (the time measured in its rest frame), and we use the Minkowski metric: $x_\mu = [x(\tau); i x_0(\tau)]$, with units in which $c = h = 1$. Note the "mass shell" property, obtained by multiplication of (4.1) by $\sum_\mu dx_\mu/d\tau$ is maintained: $\sum_\mu (dx_\mu/d\tau)^2 = \text{constant} \to -1$, since the particle's 4-momentum is given by $p_\mu = m\, dx_\mu/d\tau$, and $p^2 + m^2 = 0$.

A formal, first-integral of (4.1) can be written in terms of the OE

$$\frac{dx_\mu}{d\tau} = \left(e^{g \int_0^\tau d\tau' F(x(\tau'), \tau')} \right)_{+, \mu\nu} \cdot w_\nu, \tag{4.2}$$

where $g = e/m$ and $w_\mu = dx_\mu/d\tau|\tau = 0$. There is a considerable simplification for a laser (epw) field, where $F_{\mu\nu}(x) = f_{\mu\nu} \cos(k \cdot x)$, $f_{\mu\nu} = k_\mu \epsilon_\nu - k_\nu \epsilon_\mu$, with, $k^2 = k \cdot \epsilon = 0$, and where $k_\mu = (0, 0, \omega; i\omega)$ is the 4-momentum of the laser photons of energy ω, and ϵ_μ is their polarization. Since $(f^2)_{\mu\nu} = \sum_\lambda f_{\mu\lambda} f_{\lambda\nu} = -k_\mu k_\nu \epsilon^2$, then $(f^3)_{\mu\nu} = (f^n)_{\mu\nu} = 0$, for $n \geq 3$. In this way, the OE reduces to just three terms,

$$\delta_{\mu\nu} + g f_{\mu\nu} \int_0^\tau d\tau' \cos(k \cdot x(\tau'))$$

$$+ g^2 \int_0^\tau d\tau_1 \int_0^{\tau_1} d\tau_2 \cos(k \cdot x(\tau_1)) \cos(k \cdot x(\tau_2)) \cdot f_{\mu\lambda} f_{\lambda\nu}.$$

The ordering of the third term is trivial, and it may be rewritten as $\frac{1}{2}[\int_0^\tau d\tau' \times \cos(k \cdot x(\tau'))]^2$; and, with $u(\tau) = k \cdot x(\tau)$, one has

$$\frac{dx_\mu}{d\tau} = w_\mu + gf_{\mu\nu}w_\nu \int_0^\tau d\tau' \cos(u(\tau')) - k_\mu(k \cdot w)\frac{g^2\epsilon^2}{2}\left[\int_0^\tau d\tau' \cos(u(\tau'))\right]^2.$$

(4.3)

To determine $u(\tau)$, multiply (4.3) by $\sum_\mu k_\mu$, for which case the equation collapses to $du/d\tau = k \cdot w$, with solution $u(\tau) = \delta + \tau(k \cdot w)$, where δ is a constant. The integrals of (4.3) are then elementary, and one finds

$$\frac{dx_\mu}{d\tau} = w_\mu + \frac{gf_{\mu\nu}w_\nu}{(k \cdot w)}[\sin(\delta + \tau(k \cdot w)) - \sin\delta]$$

$$- k_\mu\frac{g^2\epsilon^2}{2(k \cdot w)}[\sin(\delta + \tau(k \cdot w)) - \sin\delta]^2. \qquad (4.4)$$

It should be noticed that $\sum_\mu(dx_\mu/d\tau)^2 = w^2$.

We choose the simplest initial conditions such that $\tau = 0$ at $x_0 = t = 0$, at which time $x_i(0) = 0$ and $dx_i/d\tau|_{\tau=0} = 0$. Since $dx_0/d\tau = dt/d\tau = E(\tau)/m$, where $E(\tau)$ is the particle's kinetic energy at proper time τ, at $\tau = 0$, $E = m$. Here, $w_\mu = (0; i)$ and $k \cdot w = -\omega$; also, $\delta = k \cdot x(0) = 0$. Extracting the time component of (4.4) then yields

$$E(\tau)/m = 1 + \frac{g^2\epsilon^2}{2}\sin^2(\omega\tau). \qquad (4.5)$$

Further, $\frac{dx_i}{d\tau} = \frac{dx_i}{dt} \cdot \frac{dt}{d\tau} \equiv v_i(\tau) \cdot \frac{E(\tau)}{m}$, so that the spatial components of the particle's velocity $v_i(\tau)$ are given by

$$v_i(\tau) = \frac{g\epsilon_i \sin(\omega\tau) + \frac{k_i}{\omega} \cdot \frac{g^2\epsilon^2}{2} \cdot \sin^2(\omega\tau)}{\left[1 + \frac{g^2\epsilon^2}{2}\sin^2(\omega\tau)\right]}. \qquad (4.6)$$

With the polarization chosen as $\epsilon_\mu = (\epsilon_1, \epsilon_2, 0; 0)$, one obtains the transverse velocities

$$v_{i=1,2} = \frac{g\epsilon_i \sin(\omega\tau)}{\left[1 + \frac{g^2\epsilon^2}{2}\sin^2(\omega\tau)\right]}, \qquad (4.7)$$

and the longitudinal velocity

$$v_3 = \frac{\frac{g^2\epsilon^2}{2}\sin^2(\omega\tau)}{\left[1 + \frac{g^2\epsilon^2}{2}\sin^2(\omega\tau)\right]}; \qquad (4.8)$$

and the requirement that $v^2 < c^2 = 1$ is clearly satisfied.

Finally, one must invert $t(\tau)$ into the form $\tau(t)$, in order to obtain expressions for $x_i(t) \equiv x_i(\tau(t))$ and $v_i(t) \equiv v_i(\tau(t))$; and for this (4.4) can be integrated once, and with $x_\mu(0) = 0$, yields

$$x^{(\tau)}_{i=1,2} = \frac{g\epsilon_i}{\omega}[1 - \cos(\omega\tau)], \qquad x_3(\tau) = \frac{g^2\epsilon^2}{4}\left[\tau - \frac{\sin(2\omega\tau)}{2\omega}\right],$$

and

$$t(\tau) = \tau\left[1 + \frac{g^2\epsilon^2}{4}\left(1 - \frac{\sin(2\omega\tau)}{2\omega\tau}\right)\right]. \tag{4.9}$$

In general, the desired inversion of (4.9) cannot be performed analytically, although approximate, or "averaged" inversions are possible. We bypass this difficulty by choosing that value of $\tau = \tau_0 = \pi/2\omega$, when $E(\tau)$ and $v_3(\tau)$ have their maximum, "first peak" values, and evaluate all quantities at that (proper) time,

$$t(\tau_0) = \frac{\pi}{2\omega}\left[1 + \frac{g^2\epsilon^2}{4}\right], \qquad x_3(\tau_0) = \frac{g^2\epsilon^2}{4}\left(\frac{\pi}{2\omega}\right), \qquad x^{(\tau_0)}_{i=1,2} = \frac{g\epsilon_i}{\omega},$$

$$v_3(\tau_0) = \frac{g^2\epsilon^2/2}{1 + g^2\epsilon^2/2}, \qquad \frac{E(\tau_0)}{m} = 1 + \frac{g^2\epsilon^2}{2}.$$

For intense lasers, with $(g\epsilon)^2 \gg 1$, one finds longitudinal velocities close to the speed of light. But if this epw-idealization of a laser beam is to be at all realistic, the extent of the particle's transverse motion cannot exceed the beam width D, so that $g\epsilon/\omega = g\epsilon\lambda/2\pi < D$. For "squeezed" lasers of $\lambda \sim 1\,\mu m$, and beam width $D \sim 10\,\mu m$, this means $(g\epsilon)^2 \lesssim 10^3$, so that the order of magnitude of the maximum KE that can be realized before the particle leaves the beam is about a thousand times its rest-mass energy.

4.2 The "scalar" laser solution for $G_c[A]$

The simplest idealized plane-wave "laser" solution occurs for the scalar G_c with scalar interaction $A(x) \Rightarrow A(k \cdot x)$, where the functional form of A is arbitrary, but $k^2 = \mathbf{k}^2 - k_0^2 = 0$; although less complicated than the full laser solution of QED, the essential features of solubility are the same.

Here, one makes the substitution in (3.14),

$$\int_0^s ds' A\left(y - \int_0^{s'} v\right) \rightarrow \int_0^s ds' A\left(k \cdot y - \int_0^{s'} ds'' k \cdot v(s'')\right), \tag{4.10}$$

and, in the manner of (3.45), inserts an expression for unity under the integrals of (3.14) with the express purpose of extracting the v-dependence which appears in the argument of A,

$$G_c(x, y|A) = i \int_0^\infty ds\, e^{-ism^2} \int \frac{d^4 p}{(2\pi)^4} e^{ip\cdot(x-y)} N' \int d[u] \int d[\Omega] e^{i\int_0^s u\Omega} \cdot \mathcal{F}[u]$$

$$\cdot e^{i\int_0^s \frac{\delta^2}{\delta v^2}} \cdot e^{i\int_0^s ds' v_\mu(s')[p_\mu - \int_{s'}^s ds'' k_\mu \Omega(s'')]}\Big|_{v\to 0}, \tag{4.11}$$

where $\mathcal{F}[u] = \exp[-ig \int_0^s ds' A(k\cdot y - u(s'))]$, and Abel's replacement of $\int_0^s ds' \Omega(s') \int_0^{s'} ds'' v_\mu(s'')$ by $\int_0^s ds' v_\mu(s') \int_{s'}^s ds'' \Omega(s'')$ has again been used. The Fradkin functional operation of the second line of (4.11) is now immediate, and yields

$$\exp\left\{ -i \int_0^s ds' \left[p_\mu - k_\mu \int_{s'}^s \Omega \right]^2 \right\} = \exp\left[-isp^2 + 2ip\cdot k \int_0^s ds' \cdot s'\Omega(s') \right], \tag{4.12}$$

where the inverse of Abel's trick has once more been used. The essential feature which guarantees solubility is then apparent: because $k^2 = 0$, there is no quadratic Ω-dependence, and its functional integral yields the simple delta functional: $\delta[u(s') + 2s' p \cdot k]$. Then, $\int d[u]$ is immediate, replacing $\mathcal{F}[u]$ by $\mathcal{F}[-2s' p \cdot k]$,

$$G_c(x, y|A) = i \int_0^\infty ds\, e^{-ism^2} \int \frac{d^4 p}{(2\pi)^4} e^{ip\cdot(x-y)} e^{-isp^2} \cdot e^{-ig\int_0^s ds' A(k\cdot y + 2s' p\cdot k)}. \tag{4.13}$$

Note that in the QED pair-production problem of Section 3.5, which restricts the vector potential to a similar argument, there is an extra term proportional to A_μ, which contains $u(s')$ dependence and appears in the corresponding delta functional, leading to a non-trivial determinantal factor.

Equation (4.13) can be further simplified. In essence, with $z = x - y$, one requires the integral

$$\int \frac{d^4 p}{(2\pi)^4} e^{ip\cdot z - isp^2} Q(k\cdot p), \tag{4.14}$$

where Q may be read off directly from (4.13). One proceeds in a manner analogous to that used above by introducing under the integrals of (4.14) a factor of unity,

$$1 = \int_{-\infty}^{+\infty} du \int_{-\infty}^{+\infty} \frac{d\omega}{2\pi} e^{i\omega(u - k\cdot p)},$$

from which one obtains

$$G_c(x, y|A) = i \int_0^\infty ds\, e^{-ism^2} \int_{-\infty}^{+\infty} du\, e^{-ig \int_0^s ds'\, A(k \cdot y - 2s'u)} \cdot \int_{-\infty}^{+\infty} \frac{d\omega}{2\pi} e^{i\omega u}$$

$$\cdot \int \frac{d^4p}{(2\pi)^4} e^{-isp^2 + ip \cdot [z - k\omega]}. \tag{4.15}$$

The last line of (4.15) is a simple Gaussian, yielding

$$\left(\frac{-i}{16\pi^2 s^2} \right) e^{i(z - k\omega)^2/4s},$$

and one notes that because $k^2 = 0$, the ω^2-term in the expansion of the exponential factor is missing, so that the ω integration generates

$$\delta(u - k \cdot z/2s),$$

permitting the u-integral to be performed. With the variable change $\lambda = s'/s$, one obtains

$$G_c(x, y|A) = \frac{1}{16\pi^2} \int_0^\infty \frac{ds}{s^2} e^{-is[m^2 + g \int_0^1 d\lambda\, A(k \cdot \xi(\lambda))] + \frac{i(x-y)^2}{4s}}, \tag{4.16}$$

where $\xi_\mu(\lambda) = \lambda x_\mu + (1 - \lambda)y_\mu$ represents the straight-line path between the points y_μ and x_μ. Finally, one realizes that (4.16) is just the ordinary, scalar, causal, Boson propagator $\Delta_c(x - y, M^2) = G_c(z)$ of (1.41) and (1.42), but with its mass^2 replaced by a position-dependent mass^2: $m^2 \to M^2 = m^2 + g \int_0^1 d\lambda\, A(k \cdot \xi(\lambda))$. A quite similar result appears for that scalar $G_c[A]$ when the argument of A is restricted to a simple, light-cone variable; and that exact Green's function solution will find application in Section 8.4.

4.3 The QED laser solutions for $G_c[A]$ and $L[A]$

Historically, solutions to this QED problem were first given by Schwinger[2] in his 1951 seminal paper on Gauge Invariance and Vacuum Polarization. One can obtain the same results starting from the Fradkin representation (3.32) for the propagator $G_c[A]$ in the presence of the external field $A_\mu(x) = \epsilon_\mu A(k \cdot x)$, where $k^2 = k \cdot \epsilon = 0$, and $A(k \cdot x)$ is arbitrary.

In order to display, subsequently, a related effect, we shall begin not with the precise form (3.32) – which would be the simplest approach – but with a $G_c[A]$ written in the form

$$G_c(x, y|A) = [m - \gamma \cdot (\partial_x - ig A(x)] \cdot \mathcal{J}_c(x, y|A), \tag{4.17}$$

where the Fradkin analysis has been carried through only for

$$\mathcal{J}_c(x, y|A) = i \int_0^\infty ds\, e^{-ism^2} \cdot e^{i\int_0^s ds'\frac{\delta^2}{\delta v^2}} \cdot \delta\left(x - y + \int_0^s v\right)$$

$$\cdot e^{-ig\int_0^s ds'\, v_\mu(s')A_\mu(y - \int_0^{s'} v)} \cdot \left(e^{g\int_0^s ds'\sigma \cdot F(y - \int_0^{s'} v)}\right)_+ \Big|_{v_\mu \to 0}. \quad (4.18)$$

One again inserts a factor of unity under the integrals of (4.18), in the form

$$N^{1} \int d[u] \int d[\Omega] e^{i\int_0^s ds'\Omega(s')[u(s') - k_\mu \int_0^{s'} ds'' v_\mu(s'')]},$$

and introduces the Fourier representation of $\delta(x - y + \int_0^s ds' v(s'))$, along with the notation $z = x - y$, $f_{\mu\nu} = k_\mu \epsilon_\nu - \epsilon_\mu k_\nu$, so that (4.18) then becomes

$$\mathcal{J}_c(x, y|A) = i \int_0^\infty ds\, e^{-ism^2} \cdot N' \int d[u] \int d[\Omega] e^{i\int_0^s ds'\Omega u}$$

$$\cdot \int \frac{d^4 p}{(2\pi)^4} e^{ip \cdot z} \cdot \left(e^{g\int_0^s ds'(\sigma \cdot f)A'(k \cdot y - u(s'))}\right)_+$$

$$\cdot e^{i\int_0^s \frac{\delta^2}{\delta v^2}} \cdot e^{i\int_0^s ds' v_\mu(s')[p_\mu - k_\mu \int_{s'}^s \Omega - g\epsilon_\mu A(k \cdot y - u(s'))]}\Big|_{v \to 0}. \quad (4.19)$$

Again, the result of the Fradkin functional operation is immediate, replacing the last line of (4.19) by

$$\exp\left\{-i\int_0^s ds'\left[p - k\int_{s'}^s \Omega - g\epsilon A(k \cdot y - u(s'))\right]^2\right\}$$

or

$$\exp\left\{-isp^2 - ig^2\epsilon^2 \int_0^s ds' A^2(k \cdot y - u(s')) + 2ip \cdot k \int_0^s ds' s'\Omega(s')\right.$$

$$\left. + 2ig\epsilon \cdot p \int_0^s ds' A(k \cdot y - u(s'))\right\}, \quad (4.20)$$

and again, the $\int d[\Omega]$ may be performed, resulting in $\delta(u(s') + 2s'p \cdot k)$, so that (4.20) becomes

$$\exp\left\{-isp^2 - ig^2\epsilon^2 \int_0^s ds' A^2(k \cdot y + 2s'p \cdot k)\right.$$

$$\left. + 2ig\epsilon \cdot p \int_0^s ds' A(k \cdot y + 2s'p \cdot k)\right\}.$$

As in the scalar computation, we are left with the integral

$$\int \frac{d^4 p}{(2\pi)^4} \mathcal{F}(k \cdot p) e^{-isp^2 + ip \cdot [z + Q(p \cdot k)]}, \quad (4.21)$$

which has the slight complication of an extra phase factor, dependent on $Q_\mu(p \cdot k) = 2g\epsilon_\mu \int_0^s ds' A(k \cdot y + 2s'p \cdot k)$. Nevertheless, (4.21) may be evaluated in the same fashion, as

$$\int_{-\infty}^{+\infty} du \int_{-\infty}^{+\infty} \frac{d\omega}{2\pi} \mathcal{F}(u) e^{iu\omega} \int \frac{d^4p}{(2\pi)^4} e^{-isp^2 + ip \cdot [z + Q(u) - \omega k]}.$$

The Gaussian momentum integral yields

$$\left(\frac{-i}{16\pi^2 s^2}\right) \exp\left\{i\frac{z^2}{4s} + i\frac{Q^2(u)}{4s} + i\frac{z \cdot Q(u)}{2s} - i\omega\frac{z \cdot k}{2s}\right\},$$

and, again, integration over ω generates $\delta(u - z \cdot k/2s)$, so that, with $\xi_\mu(\lambda) = \lambda x_\mu + (1 - \lambda)y_\mu$, one finally obtains

$$\mathcal{J}_c(x, y|A) = e^{ig\epsilon \cdot z \int_0^1 d\lambda A(k \cdot \xi(\lambda))} \cdot \left(\frac{1}{16\pi^2}\right) \int_0^\infty \frac{ds}{s^2} e^{-ism^2 + iz^2/4s}$$

$$\cdot e^{-isg^2\epsilon^2[\int_0^1 d\lambda A^2(k \cdot \xi) - (\int_0^1 d\lambda A(k \cdot \xi))^2]} \cdot e^{g(\sigma \cdot f) \int_0^1 d\lambda A'(k \cdot \xi)}. \quad (4.22)$$

In writing (4.22), the ordering symbol has been omitted because this choice of field implies $\sigma \cdot F \to (\sigma \cdot f)A'$, where $\sigma \cdot f$ is a constant matrix. In fact, this exponential factor may be further simplified because $(\sigma \cdot f)^2 = 0$; this is most simply seen by fixing the spacelike $\epsilon_\mu = (\epsilon_1, \epsilon_2, 0; i0)$, and the null $k_\mu = (0, 0, \omega; i\omega)$ so that $\sigma \cdot f = \omega(\epsilon_1\gamma_1 + \epsilon_2\gamma_2)(\gamma_3 + i\gamma_4)$, and observing that $(\gamma_3 + i\gamma_4)^2 = 0$. All terms higher than linear, in the expansion of this exponential factor, will therefore vanish.

The essential features of (4.22) are then a multiplicative phase factor,

$$e^{i\phi(x,y)} = e^{ig\epsilon \cdot z \int_0^1 d\lambda A(k \cdot \xi)},$$

and the configuration-space-dependent "variable mass" term, $M^2 = m^2 + g^2\epsilon^2\langle(\Delta A)^2\rangle$, with $\langle(\Delta A)^2\rangle = \int_0^1 d\lambda A^2(k \cdot \xi) - (\int_0^1 d\lambda A(k \cdot \xi))^2$:

$$\mathcal{J}_c(x, y|A) = e^{i\phi(x,y)}\left[1 + ig(\sigma \cdot f)\int_0^1 d\lambda A'(k \cdot \xi) \cdot \frac{\partial}{\partial m^2}\right]\Delta_c(x - y; M^2(x, y)).$$

$$(4.23)$$

The desired Green's function, $G_c(x, y|A)$, is then given by the operation of the first RHS factor of (4.17) upon the $\mathcal{J}_c(x, y|A)$ of (4.23). With the aid of the identity

$$\int_0^1 d\lambda[A(k \cdot \xi(\lambda)) + \lambda k \cdot (x - y)A'(k \cdot \xi(\lambda))]$$

$$= \int_0^1 d\lambda \frac{\partial}{\partial\lambda}[\lambda A(k \cdot \xi(\lambda))] = A(k \cdot x),$$

it is not difficult to see that $G_c(x, y|A)$ may be rewritten as

$$
G_c(x, y|A) = e^{i\phi(x,y)} \left[m + \gamma_\mu \left(\partial_\mu^x - ig(x-y)_\nu \cdot \int_0^1 d\lambda \cdot \lambda F_{\mu\nu}(k \cdot \xi(\lambda)) \right) \right]
$$
$$
\cdot \left[1 + ig\sigma_{\mu\nu} \int_0^1 d\lambda F_{\mu\nu}(k \cdot \xi(\lambda)) \frac{\partial}{\partial m^2} \right] \cdot \Delta_c(x, y|M^2(x, y)),
$$

$$(4.24)$$

which explicitly shows that the only gauge dependence of $G_c[A]$ resides in the muliplicative phase factor $i\phi(x, y)$, and is associated with gauge transformations of the form $\epsilon_\mu \to \epsilon_\mu + \alpha k_\mu$, with arbitrary α.

It is interesting to ask how M^2 can differ from m^2, and for this we consider a specific form of plane wave, $A(k \cdot x) = \cos(k \cdot x)$, and two limiting cases. (a) For $|k \cdot (x - y)| \to 0$, $k \cdot \xi$ is independent of λ, so that $\langle (\Delta A)^2 \rangle \to 0$, and $M^2 \to m^2$. The leading terms entering into this cancellation are of order ω^4, and hence sufficiently-soft laser photons do not influence charged-particle propagation. (b) For $|k \cdot (x - y)| \to \infty$, $\langle (\Delta A)^2 \rangle \to 1/2$, so that $M^2 \to m^2 + g^2\epsilon^2/2 = m^2 + g^2 U/2\omega^2$, where U is proportional to the laser energy density. This suggests that the quantum-mechanical, effective mass of a charged particle moving over considerable distances within a laser beam is increased, depending upon the intensity of the beam.

It will be useful to derive the corresponding $L[A]$, and simplest to construct it from the representations of (3.33) and (3.34). As did Schwinger, we will find that the result vanishes, a situation that can be guessed (as below) from the Physics of the situation; but it is not so much the result that is of interest here, as much as the techniques used for its extraction. In particular, it was shown in Chapter 3 that $L[A]$ is really a functional of $F_{\mu\nu}$,

$$
L[F] = -\frac{1}{2} \int_0^\infty \frac{ds}{s} e^{-ism^2} \int \frac{d^4 p}{(2\pi)^4} \int d^4 x \cdot e^{i \int_0^s \frac{\delta^2}{\delta v^2}} \cdot e^{ip \cdot \int_0^s ds' v(s')}
$$
$$
\cdot \text{tr} \left\{ \exp \left[-ig \int_0^s ds_1 v_\mu(s_1) \int_0^{s_1} ds_2 v_\nu(s_2) \int_0^1 \lambda \, d\lambda F_{\mu\nu}\left(x - \lambda \int_0^{s_1} v \right) \right. \right.
$$
$$
\left. \left. + g \int_0^s ds' \sigma \cdot F\left(x - \int_0^{s'} v \right) \right] - (g \to 0) \right\},
$$

$$(4.25)$$

which immediately generates the gauge-invariant form associated with current conservation. Because any current $\langle j_\mu(x) \rangle$ induced in the vacuum by an arbitrary source $A_\nu(y)$ satisfies the relation[3]

$$
\frac{\delta L}{\delta A_\mu(x)} = ig \langle j_\mu(x) \rangle,
$$

for that current to be conserved it must be true that $\partial_\mu^x \frac{\delta L}{\delta A_\mu(x)} = 0$. But if $L = L[F]$, as is clear from (4.25), then

$$\frac{\delta L}{\delta A_\mu(x)} = \sum_{\nu,\rho} \int d^4y \frac{\delta L}{\delta F_{\rho\nu}(y)} \frac{\delta F_{\rho\nu}(y)}{\delta A_\mu(x)} = \sum_{\nu,\rho} (\partial_\rho^x \delta_{\nu\mu} - \partial_\nu^x \delta_{\rho\mu}) \frac{\delta L}{\delta F_{\rho\nu}(x)},$$

and this current is identically conserved.

For the present choice of a laser field, and using our previous notation, one may evaluate (4.25) by again inserting the same form of unity under the integrals, to obtain

$$L[F] = -\frac{1}{2} \int_0^\infty \frac{ds}{s} e^{-ism^2} \int d^4x \cdot N^1 \int d[u] \int d[\Omega] \exp\left[i \int_0^s ds' \Omega(s') u(s')\right]$$

$$\cdot \int \frac{d^4p}{(2\pi)^4} \text{tr}\left[\left(\exp\left[g \int_0^s ds'(\sigma \cdot f) A'(k \cdot x - u(s'))\right]\right)_+\right]$$

$$\cdot \mathcal{F}[u] - \{g \to 0\}, \tag{4.26}$$

where

$$\mathcal{F}[u] = e^{i\int_0^s ds' \frac{\delta^2}{\delta v^2}} \cdot \exp\left[-i \int_0^s ds_1 \int_0^s ds_2 v_\mu(s_1) K_{\mu\nu}(s_1, s_2) v_\nu(s_2)\right] \cdot e^{-i\int_0^s v_\mu Q_\mu}, \tag{4.27}$$

with

$$K_{\mu\nu}(s_1, s_2) = g\theta(s_1 - s_2) f_{\mu\nu} \int_0^1 \lambda \, d\lambda A'(k \cdot x - u(s_1))$$

and

$$Q_\mu(s') = p_\mu - k_\mu \int_{s'}^s ds'' \Omega(s'').$$

Because $(\sigma \cdot f)^n = 0$, $n \geq 2$, and $\text{tr}(\sigma_{\mu\nu}) = 0$, the trace of the exponential $\sigma \cdot f$ term $= \text{tr}(1) = 4$, while the functional operation of (4.27) is Gaussian, and, by (2.14), yields

$$\exp\left[-i \int Q \cdot (1 + 2K)^{-1} \cdot Q - \frac{1}{2} \text{tr} \ln(1 + 2K)\right]. \tag{4.28}$$

The determinantal factor of (4.28) may be evaluated by expanding $\text{tr} \ln(1 + 2K)$ in powers of K, and noting that, because of "retardedness": $\theta(s_1 - s_2) \cdot \theta(s_2 - s_1) = 0$; the only possible non-zero contribution would be the term linear in K; but since $\text{tr}(f_{\mu\nu}) = 0$, the exponential of this factor may be replaced by unity. In the remaining Q-dependence, the requirement $k_\mu f_{\mu\nu} = 0$ removes all the K terms in the expansion of $(1 + 2K)^{-1}$, when such dependence is associated with that portion of Q_μ proportional to k_μ. In this way, (4.28) may be replaced by

$$\exp\left[-i\int\int_0^s ds_1\, ds_2\, p_\mu\langle s_1|(1+2K)_{\mu\nu}^{-1}|s_2\rangle p_\nu + 2i p\cdot k\int_0^s ds'\cdot s'\Omega(s')\right],$$

(4.29)

where we have dropped the term proportional to k^2, and used $\langle s_1|s_2\rangle = \delta(s_1 - s_2)$.

Integration over Ω again produces the factor $\delta[u(s') + 2s'p\cdot k]$, and then $\int d[u]$ generates

$$L[F] \rightarrow -2\int_0^\infty \frac{ds}{s}\, e^{-ism^2}\int\frac{d^4p}{(2\pi)^4}\int d^4x\, \bar{\mathcal{F}}(k\cdot p) - (g\rightarrow 0), \quad (4.30)$$

with

$$\bar{\mathcal{F}}(k\cdot p) = \exp\left[-i\int\int_0^s ds_1\, ds_2\, p_\mu(1+2K)_{\mu\nu}^{-1}p_\nu\right] \equiv \exp[-ip_\mu Q_{\mu\nu}p_\nu].$$

As in the preceding section, we introduce a factor of unity into the p-integrand, so that

$$L[F] \rightarrow -2\int_0^\infty \frac{ds}{s}\, e^{-ism^2}\int\int_{-\infty}^{+\infty}\frac{du\, d\omega}{2\pi}\, e^{i\omega u}\int\frac{d^4p}{(2\pi)^4}$$
$$\cdot\int d^4x\, e^{-ip_\mu Q_{\mu\nu}p_\nu - iwk\cdot p} - \{g\rightarrow 0\}. \quad (4.31)$$

The Gaussian $\int d^4p$ then produces the factor

$$-i\pi^2[\det Q]^{-1/2}\cdot\exp\left[i\frac{\omega^2}{4}k_\mu(Q^{-1})_{\mu\nu}k_\nu\right], \quad (4.32)$$

which may be most easily evaluated by adopting an expansion of Q^{-1} in powers of K,

$$(Q^{-1})_{\mu\nu} = \frac{1}{s}\left[\delta_{\mu\nu} + \frac{2}{s}\int\int_0^s ds_1\, ds_2 K_{\mu\nu}(s_1, s_2) + \cdots\right],$$

where, here,

$$K_{\mu\nu}(s_1, s_2) = 2g\int_0^1 \lambda\, d\lambda f_{\mu\nu}A'(k\cdot x + 2\lambda s_1 u)\theta(s_1 - s_2).$$

Because $k^2 = k_\mu f_{\mu\nu} = 0$, the exponential factor of (4.32) vanishes; and this then allows $\int d\omega$ to generate a $\delta(u)$, so that $\int du$ effectively removes all s_i

dependence from the argument of A' which defines K, and $Q_{\mu\nu}$ takes on the simple form,

$$Q_{\mu\nu} \to s\delta_{\mu\nu} - (2gA')\frac{s^2}{2}f_{\mu\nu} + (2gA')^2\frac{s^3}{3!}(f^2)_{\mu\nu}.$$

Finally, with the aid of the relation $\det[(Q)]^{-1/2} = \exp[-(1/2)\mathrm{tr}\,\ln(Q)]$, where tr now refers to Lorentz indices, and using the properties $\mathrm{tr}(f^n) = 0, n \geq 1$, one sees that $\det[(Q)]^{-1}$ reduces to s^{-2}, and is independent of g. This term would give a divergent contribution as $s \to 0$, except that (4.31) requires the subtraction of all $g = 0$ dependence; and hence, for this simple laser field, the resulting $L[F]$ vanishes.

This null result can be guessed by considering the simple kinematics required were $2\mathrm{Re}\,L[A]$ – which corresponds to the log of the vacuum persistence probability, as in Section 3.5 – not to vanish. Remembering that $L[A]$ is given pictorially as that virtual process consisting of the sum over all Feynman graphs corresponding to one closed fermion-loop, to which are here attached all possible numbers (starting with two) of the real laser photons, it is clear that if all of those photons are of the same 4-momentum, the absorptive part of any such loop can never correspond to the reaction $nk \to p + p'$, where n photons are selected from the beam to convert into an electron–positron pair: the (4-dimensional) square of the LHS must vanish, for real photons satisfying $k^2 = 0$, while the negative of the RHS is, in the CM system, the total e^+e^- (energy)^2. If a dispersion relation exists for this process, as one expects from a simple causality argument, where the dispersive part of the vacuum's dielectric function is given as an integral over its absorptive part, then the vanishing of the latter suggests that the dispersive part of the dielectric function is just a constant, corresponding to our result that the entire $L[A]$ must vanish.

4.4 Pair production via crossed lasers

It has been noted immediately above that pair production cannot occur in the field of a single laser, of whatever intensity, because the conservation laws $nk_\mu = p_\mu + p'_\mu$ cannot be satisfied. The situation changes qualitatively in the overlap region of two (or more) crossed lasers[4], since solutions exist for $n_1 k_\mu^{(1)} + n_2 k_\mu^{(2)} = p_\mu + p'_\mu$, for a variety of integers $n_{1,2}$. For calculational simplicity we shall assume that both lasers are composed of photons of the same energy $\simeq 1$ eV, that the lasers are oriented at a relative angle of $90°$, with a zero angle between their polarization vectors, and with an arbitrary phase difference between their fields; for purposes of estimation, we assume the lasers to have identical intensities $F_0 = 10^{22}$ Watts/meter^2, each producing a beam over a small area of dimension $D \sim 10^{-5}$ meters, with a pulsed duration of 10^2 femtoseconds (10^{-13}). These numbers define our "ideal", currently highest-intensity laser, and will be used at the end of this section when obtaining numerical estimates for the rate of such pair production. An explicit, exact solution to this problem does not seem to be possible; but the techniques of this section provide an introduction to methods which can be employed to provide estimates for such processes.

We again neglect the radiative corrections of the quantized photon field, which in the presence of an external potential A_μ^{ext} appear in the generalization of (2.40) as

$$\langle 0|S[A^{\text{ext}}]|0\rangle = e^{-\frac{i}{2}\int \frac{\delta}{\delta A} D_c \frac{\delta}{\delta A}} e^{L[A^{\text{ext}}+A]}\Big|_{A\to 0}, \qquad (4.33)$$

using instead $\langle 0|S[A^{\text{ext}}]0\rangle = \exp\{L[A^{\text{ext}}]\} = \exp[-\Gamma t/2 + i\phi]$, where one assumes that the field is turned on at time $t = 0$; and, henceforth, we suppress the superscript ext. The reason why such radiative corrections are neglected here is that any charged particle so produced will find itself in the presence of intense laser beams, and its subsequent motion may be expected to be essentially classical. This is not true for the gluon-mediated "radiative corrections" of QCD, as discussed below.

Physically, we are asking for the amplitude for a total of at least n laser photons to be absorbed coherently, and to make an e^+e^- pair produced at rest in their CM, $n = 2mc^2/\hbar\omega = 10^6$; this means a factor of g^n in the production amplitude, and a factor of α^n in the cross section. What could possibly compensate such a minuscule factor? The fact that in the overlap volume D^3 of the crossed lasers, there can be N "available" photons, and the production probability must include a counting factor similar to $N!/n! \cdot (N-n)!$, the number of ways of selecting n photons out of N available photons. If $N/n = f \gg 1$, that factor is approximately f^n; and in this way, as long as $f > \alpha^{-1}$, the multiple factors of α^n are effectively neutralized. By using a functional representation for $L[A]$, all such counting factors are automatically, and correctly, included.

We again begin with the exact relation (3.33) for $L[A]$, which, with the relation

$$e^{i\int_0^s ds' \frac{\delta^2}{\delta v^2}} e^{ip\cdot\int_0^s ds' v(s')} \mathcal{F}[v]\Big|_{v_\mu\to 0} = e^{-isp^2} e^{i\int_0^s ds' \frac{\delta^2}{\delta v^2}} \mathcal{F}[v-2p]\Big|_{v_\mu\to 0},$$

generates

$$L[A] = -\frac{1}{2}\int d^4x \int \frac{d^4p}{(2\pi)^4} \int_0^\infty \frac{ds}{s} e^{-is(m^2+p^2)} \cdot e^{i\int_0^s ds' \frac{\delta^2}{\delta v^2}}$$

$$\cdot \left\{ e^{-ig\int_0^s ds' [v_\mu(s')-2p_\mu]A_\mu(x+2s'p-\int_0^{s'} v)} \right.$$

$$\left. \cdot \text{tr}\left(e^{g\int_0^s ds' \sigma\cdot F(x+2s'p-\int_0^{s'} v)}\right)_+ - (g\to 0)\right\}\Big|_{v_\mu\to 0}. \qquad (4.34)$$

The entire problem reduces to the evaluation of the integrals of (4.34) for the case when the external potential $A_\mu(x) = \epsilon_\mu^{(1)} \sin(k^{(1)} \cdot x) + \epsilon_\mu^{(2)} \sin(k^{(2)} \cdot x + \delta)$ corresponds to the fields of a pair of intersecting laser beams of well-defined frequencies and polarizations. As noted above, we assume both beams have the same frequency ω, the same magnitudes of polarization ϵ, and the same direction of polarization $\hat{e}^{(1)} = \hat{e}^{(2)} = \hat{e}$, with $\hat{k}^{(1)} \cdot \hat{k}^{(2)} = \hat{e} \cdot \hat{k}^{(1)} = \hat{e} \cdot \hat{k}^{(2)} = 0$; also, $\epsilon_\mu \to \epsilon\hat{e}$, where $\hat{e}$ lies in the $\hat{x}$ direction, $\hat{k}^{(1)} = \hat{y}$, and $\hat{k}^{(2)} = \hat{z}$. Then, $A_\mu(x) = \epsilon_\mu[\sin\delta_1 + \sin\delta_2]$, with $\delta_1 = \omega(y - t)$, and $\delta_2 = \omega(z - t) + \delta$. Until the very last step, we shall use "natural units", with $\hbar = c = 1$; here, ω and ϵ have units of mass, and the average energy density U of each laser is given by $\epsilon^2\omega^2/8\pi$.

There are three operations which must be performed in (4.34) – the functional linkages, and the x- and p-integrations – and the complexity of the result can depend on the order in which these operations are arranged. However, it is useful to realize at once that the fundamental process we are trying to describe is such that a large number of coherent photons of energy $\omega \ll m$ are to be absorbed by the produced e^+ and e^- 4-momenta, p and p'. We therefore expect that the formulae can be well-approximated by treating the absorbed photons as "soft" compared to the lepton 4-momenta; and this naturally suggests a simplifying, no-recoil approximation, of which several are available.[5]

For our problem, perhaps the simplest such approximation is obtained by dropping the remaining v-dependence inside the arguments of the A_μ and the $F_{\mu\nu}$ of (4.34), for the function of this dependence is to produce corrections to the p, p' fermion momenta as they absorb the soft, laser photons. Hence, based on the reasonable expectation that soft corrections to hard e^+e^- momenta are irrelevant, we here perform the first simplification of the exact (4.34), replacing the latter by

$$L[A] = -\frac{1}{2} \int_0^\infty \frac{ds}{s} e^{-ism^2} \int d^4x \int \frac{d^4p}{(2\pi)^4} e^{-isp^2} \cdot e^{i\int_0^s ds' \frac{\delta^2}{\delta v^2}}$$
$$\cdot \left\{ e^{-ig\int_0^s ds'[v_\mu(s') - 2p_\mu]A_\mu(x+2s'p)} \mathrm{tr}\left(e^{g\int_0^s ds' \sigma \cdot F(x+2s'p)}\right)_+ - (g \to 0) \right\}\Big|_{v_\mu \to 0}.$$
$$(4.35)$$

In the Schwinger model, the only function of the OE is to provide a contribution to the normalization of each of the sequence of essential singularities that comprise the vacuum persistence probability P_0; those singularities arise from the functional operations upon the A_μ dependence, followed by an appropriate $\int d^4p$. In the present problem, complicated by the necessity of spatial averaging, the essential singularity will also arise from the corresponding A_μ factor, with the $\sigma \cdot F$ term contributing to the normalization. Since we are only interested in the order of magnitude of Γ, generated by the essential singularity, and since we have every confidence that a complete calculation which includes the $\sigma \cdot F$ term will provide a positive Γ (that is, a negative log Re $L[A]$), we shall simply drop the $\sigma \cdot F$ OE, replacing its trace by $+4$. In principle, the entire analysis can be organized without this approximation; but it would add nothing but complexity to the extraction of the essential singularity. Thus, we further simplify the expression for $L[A]$ as

$$L[A] \rightarrow -2 \int_0^\infty \frac{ds}{s} e^{-ism^2} \int d^4x \int \frac{d^4p}{(2\pi)^2} e^{-isp^2} \cdot e^{i\int_0^s ds' \frac{\delta^2}{\delta v^2}}$$
$$\cdot \left\{ e^{-ig\int_0^s ds'[v_\mu(s')-2p_\mu]A_\mu(x+2s'p)} - 1 \right\}_{v_\mu \to 0}, \qquad (4.36)$$

and we now briefly sketch, in rapid sequence, three models for its estimation. For all details, the reader is referred to the original paper.[4]

In the First Cumulant Model, the linkage operation is first carried through, and then integration over $\int d^4p$ is attempted (by the insertion of a useful expression for unity in the integrand), resulting in a reduction to a two-fold integral over an exponential factor that contains all the configuration-space dependence. One must then integrate, or average, that spatial dependence over the overlap region of the two laser beams. We assume that the linear dimension D of this volume D^3 is significantly larger (at least by a factor of 10) than the laser wavelength $\lambda\gamma$, so that the averaging procedure adopted here and in the two subsequent models is sensible.

The "first cumulant" approximation replaces the spatial average of the exponential factor, $D^{-3} \int d^3x \exp[S(x)]$, by the exponential of the average: $\exp\{D^{-3} \int d^3x S(x)\}$. This is a familiar approximation, perhaps the simplest of those used in statistical problems to treat a full cluster expansion; in that context, it is equivalent to retaining only the Q_1 term of the expansion of (2.39). In the Second Model, the spatial averaging and a portion of $\int d^4p$ are performed exactly, but a variant of the "first cumulant" approximation is used to estimate the forms obtained. The Third Model is obtained by summing an approximation to (what are expected to be the major) contributions from each of the cluster coefficients neglected in the Second Model, and generates the most interesting result of all three estimates.

In each model, we find a result dependent upon two dimensionless variables, $g\epsilon/m$ and $\pm m/\omega$, with the latter appearing as integration limits. Since $m/\omega \sim 10^6$, we make the further approximation of replacing m/ω by ∞, in which case certain simplifications appear, and generate for the First Model the result

$$L_1[A] \rightarrow i\frac{D^3ct}{2(2\pi)^2}m^4 \int_0^\infty \frac{dt}{t^3} e^{-it} \left[\frac{1}{\sqrt{1-(\gamma t)^4}} - 1 \right], \qquad (4.37)$$

where $\gamma = g\epsilon\omega/m^2\sqrt{6}$. Thus, in the limit of arbitrarily large m/ω, but fixed $g\epsilon\omega/m^2$, $L_1 \to L_1(\gamma)$. The same behavior will occur in all three models, and its relation to the Schwinger constant-field result will be discussed below.

One now sees that an (improper) expansion of the square root of (4.37) in powers of γ will generate a sequence of imaginary contributions to L_1, for the integral $\int_0^\infty dt\, t^{1+4n} e^{it}$ is real for every integer n. Hence, $\mathrm{Re}\,L_1$ does not have an expansion in powers of the coupling; it is intrinsically non-perturbative. Its value is most easily obtained by remembering that the path of the t-integration of the original Schwinger/Fradkin representation may be taken as running just below the positive t-axis; and because the cut structure of $[1 - (t/t_0)^4]^{-1/2}$, with $t_0 = 1/\gamma$, shows that a rotation of the t-contour to run according as $t \to \epsilon - i\tau$ is permissible, (4.37) may be rewritten as

$$L_1[A] = -\mathrm{i}\frac{D^3 ct}{2(2\pi)^2}m^4 \int_0^\infty \frac{d\tau}{\tau^3}\mathrm{e}^{-\tau}\left[\frac{1}{\sqrt{1 - (\tau/t_0)^4}} - 1\right],$$

and

$$\mathrm{Re}\,L_1[A] = -\frac{D^3 ct}{2(2\pi)^2}m^4 \int_{t_0}^\infty \frac{d\tau}{\tau^3}\frac{\mathrm{e}^{-\tau}}{\sqrt{(\tau/t_0)^4 - 1}}, \qquad (4.38)$$

where the branch of the square root has been chosen to yield a negative value for $\mathrm{Re}\,L[A]$. Under the variable change $y = \tau/t_0 - 1$, the integral of (4.38) becomes

$$t_0^{-2}\mathrm{e}^{-t_0}\int_0^\infty \frac{dy}{(1+y)^3}\frac{\mathrm{e}^{-yt_0}}{\sqrt{(1+y)^4 - 1}} \simeq t_0^{-2}\frac{\mathrm{e}^{-t_0}}{2}\int_0^{t_0^{-1}} \frac{dy}{\sqrt{y}} = t_0^{-3}\mathrm{e}^{-t_0},$$

since the y-integral effectively cuts off at $t_0^{-1} \ll 1$. Hence

$$\mathrm{Re}\,L_1[A] \simeq -\frac{D^3 ct}{2(2\pi)^2}m^4\gamma^3\mathrm{e}^{-1/\gamma}, \qquad (4.39)$$

which clearly displays the essential singularity in γ.

The Second and Third Models are obtained by first performing the spatial integrations of (4.36), which we write in the form of averages over the arguments of the sin terms there, in the form

$$\int d^3x \to D^3\left(\frac{1}{D^3}\int dx \int dy \int dz\right) \to D^3\langle\cdots\rangle_{\theta,\bar{\theta}},$$

where the symbol $\langle\cdots\rangle_{\theta,\bar{\theta}}$ signifies independent averages over the factors $\sin(\theta + u(s'))$ and $\sin(\bar{\theta} + \bar{u}(s'))$, where $\theta = \omega(y - t)$, $\bar{\theta} = \omega(z - t) + \delta$, $u(s') = 2s'k^{(1)}\cdot p$, $\bar{u}(s') = 2s'k^{(2)}\cdot p$.

Upon expansion of the exponential, one can then show that the averages

$$\frac{1}{(2\pi)^2} \int_0^{2\pi} d\theta \int_0^{2\pi} d\bar{\theta}$$

$$\cdot \exp\left\{-ig \int_0^s ds' V_\mu(s') [\epsilon_\mu^{(1)} \sin(\theta + u(s')) + \epsilon_\mu^{(2)} \sin(\bar{\theta} + \bar{u}(s'))]\right\}$$

can be expressed as

$$\sum_{l=0}^{\infty} (-)^l \frac{(g^2 \frac{A_1}{2})^l}{(l!)^2} \sum_{n=0}^{\infty} (-)^n \frac{(g^2 \frac{A_2}{2})^n}{(n!)^2} = J_0\left(g\sqrt{\frac{A_1}{2}}\right) J_0\left(g\sqrt{\frac{A_2}{2}}\right), \quad (4.40)$$

where

$$A_1 = \int_0^s ds_a \int_0^s ds_b \left(V(s_a) \cdot \epsilon^{(1)}\right) \cos(u(s_a) - u(s_b)) \left(V(s_b) \cdot \epsilon^{(1)}\right),$$

$$A_2 = \int_0^s ds_a \int_0^s ds_b \left(V(s_a) \cdot \epsilon^{(2)}\right) \cos(u(s_a) - u(s_b)) \left(V(s_b) \cdot \epsilon^{(2)}\right),$$

and $V_\mu = v_\mu - 2p$.

To perform the linkage operation, it is most convenient to introduce the representation

$$J_0(z) = \frac{1}{2\pi i} \int_{-\infty}^{0+} \frac{dt}{t} e^{t - z^2/4t},$$

where the contour is specified as approaching the origin from $-\infty$ underneath the negative t-axis, swinging in a half-circle around the origin, and moving out to $-\infty$ above the negative t-axis. The Fradkin linkages are Gaussian, and yield

$$L_2[A] = -2 \frac{(D^3 ct)}{(2\pi)^4} \left(\frac{1}{2\pi i}\right)^2 \int_{-\infty}^{0+} \frac{dt_1}{t_1} e^{t_1} \int_{-\infty}^{0+} \frac{dt_2}{t_2} e^{t_2} \int_0^\infty \frac{ds}{s} e^{-ism^2}$$

$$\cdot \int d^4 p \, e^{-isp^2} \left\{ e^{-\frac{1}{2} \operatorname{Tr} \ln(1+2K)} \cdot e^{ip \cdot Q \cdot p} - 1 \right\}, \quad (4.41)$$

where

$$Q_{\mu\nu}(s) = \int_0^s ds_1 \int_0^s ds_2 \langle s_1 | \left(2K \frac{1}{1+2K}\right)_{\mu\nu} |s_2\rangle$$

and

$$\langle s_1 | K_{\mu\nu} | s_2 \rangle = i \frac{g^2}{4} \epsilon_\mu \epsilon_\nu \left[\frac{\cos\left(2p \cdot k^{(1)}(s_1 - s_2)\right)}{t_1} + \frac{\cos\left(2p \cdot k^{(2)}(s_1 - s_2)\right)}{t_2} \right].$$

Here, both K and Q depend on p, and so we introduce

$$\mathcal{F}\left(2p\cdot k^{(1)}, 2p\cdot k^{(2)}\right) = \int\int_{-\infty}^{+\infty} du_1\, du_2 \mathcal{F}(u_1, u_2) \cdot \left(\frac{1}{2\pi}\right)^2 \int\int_{-\infty}^{+\infty} d\Omega_1\, d\Omega_2$$

$$\cdot\, e^{i(u_1\Omega_1 + u_2\Omega_2) - 2ip\cdot k^{(1)}\Omega_1 - 2ip\cdot k^{(2)}\Omega_2},$$

where $\mathcal{F}(u_1, u_2)$ represents all the $2p\cdot k^{(1,2)}$ dependence in the curly bracket of (4.41). Integrating over p, and then over $\Omega_{1,2}$, one obtains

$$L_2[A] = i\frac{(D^3 ct)}{16\pi^3}\left(\frac{1}{2\pi i}\right)^2 \int_{-\infty}^{0^+}\frac{dt_1}{t_1}\, e^{t_1}$$

$$\cdot \int_{-\infty}^{0^+}\frac{dt_2}{t_2}\, e^{t_2}\int_0^\infty \frac{ds}{s^3}\, e^{-ism^2}\cdot\left(\frac{s}{2w^2}\right)\cdot\int\int_{-\infty}^{+\infty} du_1\, du_2$$

$$\cdot\, e^{iu_1 u_2 s/2w^2}\left\{e^{-\frac{1}{2}\mathrm{Tr}\ln(1+2k)}\cdot e^{-\frac{1}{2}\mathrm{tr}\ln(1-Q/s)-1}\right\}, \qquad (4.42)$$

with

$$\langle s_1|2K_{\mu\nu}(u_1, u_2)|s_2\rangle$$

$$= i\frac{g^2\epsilon^2}{2}\left(\frac{\epsilon_\mu\epsilon_\nu}{\epsilon^2}\right)\left[\frac{1}{t_1}\cos(u_1(s_1 - s_2)) + \frac{1}{t_2}\cos(u_2(s_1 - s_2))\right]$$

$$\equiv \left(\frac{\epsilon_\mu\epsilon_\nu}{\epsilon^2}\right)2K(s_1, s_2).$$

It then follows that $Q_{\mu\nu}/s$ may be written as $q(s)(\frac{\epsilon_\mu\epsilon_\nu}{\epsilon^2})$, and we henceforth suppress the factors $(\frac{\epsilon_\mu\epsilon_\nu}{\epsilon^2})$.

The calculation now reduces to the evaluation of repeated integrals over $2K(s_1, s_2)$ – we here emphasize its $s_{1,2}$ variables – e.g.,

$$\mathrm{Tr}\ln(1 + 2K) = \int_0^s ds_1\left[2K(s_1, s_1) - \frac{1}{2}\int_0^s ds_2\, 2K(s_1, s_2)\cdot 2K(s_2, s_1) + \cdots\right],$$

and adopt the notation $K(s_1, s_2) = K(s_1 - s_2)$. Note that $K(s_1, s_1) = K(0)$, which quantity would be the only one appearing were $\omega \to 0$. But, physically, for $\omega \to 0$ at fixed ϵ, L must vanish; and one must therefore find that the curly bracket of (4.42) must vanish when each $K(s_1 - s_2)$ is replaced by $K(0)$. This suggests expanding $\mathrm{Tr}\ln(1 + 2K)$ and $\mathrm{tr}(1 - q(s))$ in powers of the relevant quantity: $\delta K(s_1 - s_2) = K(0) - K(s_1 - s_2)$, and retaining only terms to first order – which defines the Second Model – and leads, after some further manipulations, to

$$\text{Re}\, L_2 \rightarrow \frac{m^4}{(4\pi)^2}(D^3 ct) \int_0^\infty \frac{d\tau}{\tau^3} e^{-\tau} \cos\tau\left[1 - J_0^2((\gamma\tau)^2)\right]. \quad (4.43)$$

One can see, directly from the perturbation expansion of (4.43), that the coefficient of every term (proportional to g^{4n}) is identically zero, so that the non-perturbative aspect of Re L has been maintained.

We are only able to evaluate (4.43) in an approximate way, and observe that the sign of our approximate result,

$$\text{Re}\, L_2 \simeq \frac{m^4}{(4\pi)^2}(D^3 ct) \cdot \frac{\gamma^3}{2} e^{-1/\gamma}[\cos(\gamma^{-1}) - \sin(\gamma^{-1})], \quad (4.44)$$

is apparently related to its oscillatory dependence. One can invent an averaging process so that the final sign is negative; but the proper answer to this question must wait upon the evaluation of the neglected $\sigma \cdot F$ terms, and then – if there is any choice of branch – to the physical requirement that the final sign of Re L must be negative. We therefore write

$$\Gamma_2 \simeq \frac{m^4(D^3 c)}{(4\pi)^2}\gamma^3 e^{-\frac{1}{\gamma}} \quad (4.45)$$

and again emphasize that it is only the order of magnitude of this result which is believed to be a correct prediction of the Second Model. Comparison with (4.39) shows that this is essentially the same result as obtained for the First Model.

The Third Model is obtained by returning to (4.42) and developing a better approximation for the combination $\text{Tr}\ln(1 + 2K) + \text{tr}\ln(1 - Q/s)$, an approximation which contains all powers of the coupling, rather than just its quadratic dependence. Further, any such realization must be simple enough to permit its evaluation in closed form.

We begin by calculating $\langle s_1|(2K)^2|s_2\rangle$, a straightforward computation; with $\chi = ig^2\epsilon^2/2$, and $s_{12} = s_1 - s_2$, one finds:

$$\langle s_1|(2K)^2|s_2\rangle = \chi^2\left\{\frac{s}{t_1^2}\left[\cos(u_1 s_{12}) + \frac{\sin(u_1 s)}{(u_1 s)}\right] + \frac{s}{t_2^2}\left[\cos(u_2 s_{12}) + \frac{\sin(u_2 s)}{(u_2 s)}\right]\right.$$
$$+ \frac{s}{t_1 t_2}\left[\cos\left(\left[\frac{u_1 + u_2}{2}\right]s_{12}\right) \cdot \frac{\sin\left(\left[\frac{u_1 - u_2}{2}\right]s\right)}{\left[\frac{u_1 - u_2}{2}\right]s}\right.$$
$$\left.\left. + \cos\left(\left[\frac{u_1 - u_2}{2}\right]s_{12}\right) \cdot \frac{\sin\left(\left[\frac{u_1 + u_2}{2}\right]s\right)}{\left[\frac{u_1 + u_2}{2}\right]s}\right]\right\}. \quad (4.46)$$

In comparison, $\langle s_1|2K|s_2\rangle = \chi[\frac{1}{t_1}\cos(u_1 s_{12}) + \frac{1}{t_2}\cos(u_2 s_{12})]$, where the trivial $\epsilon_\mu \epsilon_\nu / \epsilon^2$ factors have been suppressed. Upon subsequent integration of (4.46), the terms $\sin([\frac{u_1 \pm u_2}{2}]s)/[\frac{u_1 \pm u_2}{2}]s$ will take appreciable values only for $u_1 \pm u_2 \simeq 0$, where they become unity, so that the $\cos([\frac{u_1 \pm u_2}{2}] \cdot s_{12})$ factors multiplying them can be replaced by $\cos(u_1 s_{12})$ or $\cos(u_2 s_{12})$, precisely the terms which appear in $\langle s_1|2K|s_2\rangle$. This suggests that a model be defined by the statement

$$\langle s_1|(2K)^2|s_2\rangle = \mathcal{F}\langle s_1|2K|s_2\rangle, \tag{4.47}$$

with $\mathcal{F} = s\chi[\frac{1}{t_1} + \frac{1}{t_2}]$ (which quantity was previously called $sK(0)$). This model neglects certain oscillatory terms of (4.46), but it does correspond to an order-by-order extraction of what are probably the most significant pieces of every perturbative term, those which are not expected to vanish upon subsequent integration over fluctuating $u_{1,2}$ dependence.

With (4.47), and the arbitrary number of iterations that can be formed from it,

$$\langle s_1|(2K)^n|s_2\rangle = \mathcal{F}^{n-1}\langle s_1|2K|s_2\rangle,$$

one easily calculates

$$\mathrm{Tr}\ln(1+2K) = \ln(1+\mathcal{F})$$

and

$$\mathrm{tr}\ln(1-Q/s) = -\ln(1+\mathcal{F}) + \ln\left[1+s\chi\left(\frac{1}{t_1}\phi(u_1 s) + \frac{1}{t_2}\phi(u_2 s)\right)\right],$$

where $\phi(x) = 1 - \frac{\sin x}{x}$, so that the combination $e^{-\frac{1}{2}\mathrm{Tr}\ln(1+2K)-\frac{1}{2}\mathrm{tr}\ln(1-Q/s)} - 1$ becomes

$$\left[1 + \chi s\left(\frac{1}{t_1}\phi(u_1 s) + \frac{1}{t_2}\phi(u_2 s)\right)\right]^{-\frac{1}{2}} - 1,$$

which can be rewritten as

$$\frac{2}{\sqrt{\pi}}\int_0^\infty du\, e^{-u^2}\left\{e^{-\chi s u^2[\frac{1}{t_1}\phi(u_1 s)+\frac{1}{t_2}\phi(u_2 s)]} - 1\right\}.$$

Again, one introduces the J_0 representation following (4.40), and finds that the only difference between L_2 and this L_3 is that the latter is given by a Gaussian-weighted integral over the previous L_2, so that one can write the output of the Third Model as

$$\Gamma_3(\gamma) = \sqrt{\frac{2}{\pi}}\int_0^\infty du\, e^{-u^2/2}\Gamma_2(u\gamma),$$

or

$$\Gamma_3 = \frac{m^4(D^3c)}{(4\pi)^2}\gamma^3\sqrt{\frac{2}{\pi}}\int_0^\infty du\, e^{-u^2/2}\cdot u^3\cdot e^{-1/u\gamma}.$$ (4.48)

Since $\gamma \ll 1$, (4.48) is easily approximated as

$$\Gamma_3(\gamma) \simeq \frac{2}{\sqrt{3}}\frac{m^4(D^3c)}{(4\pi)^2}\gamma^2 e^{-3/2\gamma^{2/3}},$$ (4.49)

which represents a significant change in result compared to the previous models, mostly because the essential singularity has been changed (weakened) to $\exp[-3\gamma^{-2/3}/2]$.

This change in form of the essential singularity may have a physical interpretation which is of some interest. Elementary QED processes, such as e^+e^- pair creation, are usually thought of as taking place over distances on the order of λ_c, the electron's Compton wavelength. Here, however, we expect coherent absorption of the laser photons by the incipient, still virtual pair, over distances larger than λ_c, perhaps as large as some fraction of the laser photons' wavelength λ_γ, because there are so many photons which must be absorbed. This coherence is made explicit by the u-integration, as the parameter u varies over distances centered about $u_0 \sim \gamma^{-1/3}$, which is considerably larger than one. In physical terms, the $\exp[-1/\gamma] = \exp[-\sqrt{6}(\frac{m}{g\epsilon})(\frac{\lambda_\gamma}{\lambda_c})]$ of Γ_2 is here replaced by $\exp[-\sqrt{6}(\frac{m}{g\epsilon})(\frac{\lambda_\gamma}{u_0\lambda_c})]$, where $u_0\lambda_c$ is perhaps 10^2–10^3 times larger than λ_c, and which can be interpreted as the qualitative distance over which coherent absorption takes place.

After all the preceding analysis, it may be somewhat discouraging to learn that the intensity of even our "ideal" crossed lasers is too small by about seven orders-of-magnitude to allow a pair to be produced (assuming that a pair is to be produced after every ten pulses each of duration 10^{-13} seconds). This best estimate arises from the Third Model; and the numerical estimates may be found in the original paper.[4] But, since laser intensities appear to increase by at least an order-of-magnitude each year, it may be that the needed intensities will be available in the near future. This is a practical problem, about which the author can make no statement.

But let us suppose that sufficiently high-intensity lasers can be made which achieve e^+e^- and even $\mu^+\mu^-$ pair production; then, there is no reason why one cannot contemplate laser-induced quark–antiquark production.[6] Here one cannot neglect the QCD radiative corrections, since it is the gluon clouds surrounding q and $\bar{q}$ which form a flux tube/string, and produce quark confinement. But one can idealize what may happen in terms of two extreme, and differing, possibilities: (1) q and $\bar{q}$ appear with their flux tube/string in place, so that one has produced, in effect, a π^0, which the laser fields are incapable of tearing apart; or (2) q and $\bar{q}$ materialize, each surrounded by its virtual gluonic structure, which immediately begins to form itself into a tube/string joining q and $\bar{q}$. The formation of the tube/string is surely not an instantaneous effect, but one which can be characterized by a "string-formation velocity" v_f. As a physical process, one expects that v_f cannot be larger than c, while it is perfectly possible for the q and $\bar{q}$ to be accelerated away from each other by the crossed lasers so that their relative velocity of separation could equal or exceed v_f. This suggests that, by this mechanism, q and $\bar{q}$ might temporarily reach separation distances considerably larger than a few fermis. (Of course, after the laser beams' phases pass over the q and $\bar{q}$, deceleration occurs, and the tube/string wins.)

What could be a signal of this second possibility? Large energy deposition in a small spatial region, perhaps leading to a pair of hadronic jets, built around the outgoing q and $\bar{q}$, and arranged so as to maintain an overall color-singlet property. Other structures are also possible, such as the q and $\bar{q}$ falling back together and annihilating à la positronium, but with a relatively large energy (absorbed from the intersecting lasers when the beams pulled the q and $\bar{q}$ apart, and converted into potential energy of $q\bar{q}$ separation) converted into a few high-energy gammas, or into a "fireball" of X-rays. Much more theoretical work needs to be done on these questions; but the qualitative way in which the $q\bar{q}$ pair materialize should be amenable to experimental determination.

Notes

1 W. B. Mori and T. Katsouleas, "Laser acceleration", in the *American Institute of Physics Conference Proceedings* No. 335 (1995). A detailed discussion of electromagnetic and gravitational lasers, together with heat kernels à la Fradkin, and ordered exponentials, appears in the Ph.D thesis by P. Stojkov, "Lasers, heat kernels, and ordered exponentials", Physics Department, Brown University, 1998.

2 J. Schwinger, *Phys. Rev.* **82** (1951) 664. The interaction of a free electron with a laser beam was treated by L. S. Brown and T. W. B. Kibble, *Phys. Rev.* **133** (1964) 705, in whose paper can be found numerous references to then-current work, in particular, that of Z. Fried, *Phys. Rev. Letters* **3** (1963) 349. Subsequently, many authors have made modest extensions of the Schwinger solution by combining oscillating, epw, and constant fields in various combinations and numbers of spatial dimensions; a recent publication, which refers to many of these papers is: M. N. Hounkonnou and M. Naciri, *J. Phys. G: Nucl. Part. Phys.* **26** (2000) 1849.

3 This, among many other relations discussing current conservation and gauge invariance, was first written by Schwinger, in his many Lecture Notes and papers, and appears, e.g., in the second paper of Note 2. A detailed discussion using the notation of the present volume was given in the book by the present author, *Functional Methods and Models in Quantum Field Theory*, MIT Press, Cambridge, MA (1972).

4 H. M. Fried, Y. Gabellini, B. H. J. McKellar, and J. Avan, *Phys. Rev.* **D 63** (2001) 125001.

5 See, for example, the book by H. M. Fried, *Functional Methods and Eikonal Models*, Éditions Frontières, Gif-sur-Yvette, France (1990), as well as Chapters 5 and 7 of the present book.

6 Similar QCD questions in the context of Deep Inelastic Scattering have been treated by E. A. Paschos, *Phys. Letters B* **389** (1996) 383, which contains other references on the excitation of the QCD vacuum.

5

Special variants of the Fradkin representation

5.1 Exact representations for scalar interactions

Exact Fradkin representations have found explicit realizations in special, soluble situations of the previous two chapters. For interactions which are not soluble, one would like to have a general method for displaying the content of the exact representation for arbitrary potentials $A(z)$, which method may even suggest a form of non-perturbative approximation useful in various situations. One feature of the method presented here[1] is that it provides a qualitative measure of error for the approximations it suggests, which is at least a partial improvement over the more usual practice of generating uncontrolled approximations to these, and other, nonlinear problems. For simplicity, we begin with a scalar-potential interaction, in a relativistic framework.

These variants of the exact Fradkin representation are an outgrowth of older, no-recoil models for $G_c[A]$,[2] which will be derived and employed for specific scattering processes in Chapter 7, and which generate the so-called Block–Nordsieck (BN) functional,

$$\langle p|G_{BN}[A]|p'\rangle = i \int_0^\infty ds\, e^{-is(m^2+p^2)} \int \frac{d^4z}{(2\pi)^4}\, e^{-iq\cdot z} \cdot e^{-ig\int_0^s ds' A(z+2s'p)},$$

(5.1)

where $q = p - p'$. In any expansion of the exact $G_c[A]$ in powers of A, one finds a sequence of propagators of momentum-space form $[m^2 + (p - \sum_j k_j)^2]^{-1}$, which, in a no-recoil approximation, are replaced by $[m^2 + p^2 - 2p \cdot \sum_j k_j]^{-1}$, under the assumption that the Fourier-transform momenta k_μ of $\tilde A(k)$ are to be treated as much smaller than the momentum p_μ of the scattering particle. (More precisely, $|k_i \cdot k_j| \ll |p \cdot k_i|$ or $|p \cdot k_j|$.) This approximation, made in every order, and then summed over all orders, leads directly to the $G_{BN}[A]$ of (5.1). While most useful in problems in which sums over all

75

soft-photon contributions and processes are desired, as detailed in Chapter 7, this approximation is not without a serious drawback. The reason is that, in configuration space, the exact (scalar-interaction) $G_c(x, y|A)$ must be symmetric in x and y, while, in momentum space (with p the momentum entering and p' the momentum leaving any Feynman graph) the corresponding symmetry is $\langle p|G_c[A]|p'\rangle = \langle -p'|G_c[A]| - p\rangle$. In non-relativistic, spinless, time-independent, potential-theory scattering, this is just a statement of time-reversal invariance (TRI), which appellation we shall continue to use for the present case.

A cursory inspection of the $G_{BN}[A]$ of (5.1) shows that TRI is not satisfied by this functional; but it is satisfied by the symmetrized form:

$$\langle p|G_c^{\langle ph\rangle}[A]|p'\rangle = i \int_0^\infty ds\, e^{-is[m^2+p^2]} \int \frac{d^4z}{(2\pi)^4} e^{-iq\cdot z}$$

$$\cdot \exp\left[-ig \int_0^s ds'A(z+s'(p+p'))\right], \qquad (5.2)$$

of the approximation, which will be denoted by the superscript $\langle ph\rangle$, for "phase-averaged". (One need perform only the pair of variable changes: $z \to z - s(p + p')$, followed by $s' \to s - s'$, to demonstrate this invariance explicitly.) The approximation of (5.2) reduces to that of (5.1) in the special case when all the k of $A(k)$ are much smaller than p or p' (or, equivalently, when the magnitude of momentum transfer $|q| \ll |p|$ or $|p'|$); but it is not derived with any reference to a no-recoil approximation, and it even provides a reasonable approximation to a potential-theory scattering problem when one of the scatterings is "hard" and all the others are "soft". Clearly, $G_c^{\langle ph\rangle}[A]$ represents one further step on the road to a better approximation of $G_c[A]$ than does $G_{BN}[A]$; and the basic question of this chapter is how to define such improvement in a systematic way.

We begin with the exact Fradkin representation for the configuration-space causal propagator in four dimensions, (3.14), and expand in powers of g to obtain

$$G_c(x, y|A) = i \int_0^\infty ds\, e^{-ism^2} \int \frac{d^4p}{(2\pi)^4} e^{ip\cdot(x-y)} \sum_{n=0}^\infty \frac{(-ig)^n}{n!} \int_0^s ds_1 \cdots \int_0^s ds_n$$

$$\cdot \int \frac{d^4k_1}{(2\pi)^4} \tilde{A}(k_1) \cdots \int \frac{d^4k_n}{(2\pi)^4} \tilde{A}(k_n) \cdot \exp\left[i\sum_{l=1}^n k_l \cdot y\right]$$

$$\cdot e^{i\int_0^s ds' \frac{\delta^2}{\delta v^2}} \cdot \exp\left[i\int_0^s ds'v(s') \cdot \left[p - \sum_{l=1}^n k_l\theta(s_l - s')\right]\right]\Bigg|_{v\to 0} .$$

$$(5.3)$$

The functional operation of the last line of (5.3) is immediate, yielding

$$\exp\left\{-i\int_0^s ds'\left[p - \sum_{l=1}^n k_l\theta(s_l - s')\right]^2\right\}$$

$$= \exp\left\{-isp^2 + 2ip\cdot\sum_{l=1}^n k_l s_l - i\sum_{l,m=1}^n k_l\cdot k_m h(s_l, s_m)\right\}, \quad (5.4)$$

with

$$h(s_l, s_m) = \int_0^s ds'\theta(s_l - s')\theta(s_m - s') = \frac{1}{2}[(s_l + s_m) - |s_l - s_m|]. \quad (5.5)$$

The essential ⟨ph⟩ approximation of the reference in Note 1 was to retain the first, and simplest part of $h(s_l, s_m)$, approximating the latter by $(1/2)(s_l + s_m)$; this generates a simple "factorization" of the s-dependence such that all terms of the expansion are easily summed, generating (5.2).

In order to retain the second, and more complicated part of the $h(s_l, s_m)$ of (5.5), one needs to find a sufficiently simple representation of the remaining $-|s_l - s_m|/2$ term. Because each s_l lies in the range $0 \le s_l \le s$, the quantity $|x| = |s_l - s_m|/s$ satisfies $0 \le |x| \le 1$; and one can invoke the well-known "saw-tooth", Fourier series representation:

$$|x| = \frac{1}{2} - \frac{4}{\pi^2}\sum_{N=1,3,\ldots}^{\infty}\frac{\cos(N\pi x)}{N^2}, \quad (5.6)$$

or, since $\sum_N' \frac{1}{N^2} = \pi^2/8$, with $\sum_N' \equiv \sum_{N=1,3,5,\ldots}^{\infty}$,

$$|x| = \frac{4}{\pi^2}\sum_N' \frac{1}{N^2}[1 - \cos(N\pi x)], \quad (5.7)$$

which extends over an arbitrary number of quadrants. In our case, we need to describe $|x|$ only in the region $0 \le |x| \le 1$, which is automatically enforced by the integrals $\int_0^s ds_l$. Equation (5.6) then provides a representation for $|s_l - s_m|$ which can be put into a "factorized" form,

$$|s_l - s_m| = \frac{s}{2} - \frac{4s}{\pi^2}\sum_N' \frac{1}{N^2}\left[\cos\left(\frac{N\pi s_l}{s}\right)\cos\left(\frac{N\pi s_m}{s}\right)\right.$$

$$\left. + \sin\left(\frac{N\pi s_l}{s}\right)\sin\left(\frac{N\pi s_m}{s}\right)\right]. \quad (5.8)$$

The contribution of the third phase factor of (5.4) may then be written as

$$-i\left[\sum_l k_l s_l\right]\cdot\left[\sum_m k_m\right] + \frac{is}{4}\left[\sum_l k_l\right]^2 - \frac{2is}{\pi^2}\sum_N' \frac{1}{N^2}\left[\sum_l k_l \cos\left(\frac{N\pi s_l}{s}\right)\right]^2$$

$$- \frac{2is}{\pi^2}\sum_N' \frac{1}{N^2}\left[\sum_l k_l \sin\left(\frac{N\pi s_l}{s}\right)\right]^2. \quad (5.9)$$

It is now convenient to introduce the variables $K = \sum_l k_l$, $z = \sum_l k_l s_l$, $C_n = \sum_l k_l \cos(\frac{N\pi s_l}{s})$, $S_N = \sum_l k_l \sin(\frac{N\pi s_l}{s})$, and to rewrite the exponential of (5.9) under its $\int_0^s ds_l$ integrals in the form

$$
\int \frac{d^4K d^4Q}{(2\pi)^4} \exp\left[iQ \cdot \left(K - \sum_l k_l \right) + \frac{is}{4} K^2 \right]
$$

$$
\cdot \int \frac{d^4z d^4P}{(2\pi)^4} \exp\left[iP \cdot \left(z - \sum_l k_l s_l \right) - iz \cdot K \right]
$$

$$
\cdot \frac{\Pi'}{N} \int \frac{d^4 C_N\, d^4 P_N}{(2\pi)^4} \exp\left[iP_N \cdot \left(C_N - \sum_l k_l \cos\left(\frac{N\pi s_l}{s}\right) \right) - \frac{2is}{\pi^2}\frac{1}{N^2} C_N^2 \right]
$$

$$
\cdot \frac{\Pi'}{N} \int \frac{d^4 S_N\, d^4 Q_N}{(2\pi)^4} \exp\left[iQ_N \cdot \left(S_N - \sum_l k_l \sin\left(\frac{N\pi s_l}{s}\right) \right) - \frac{2is}{\pi^2}\frac{1}{N^2} S_N^2 \right].
$$

$$(5.10)$$

No perturbative index "n" is needed for any of these auxiliary variables K, $Q, z, P, C_N, P_N, S_N, Q_N$, for exactly the same integrals, and their weights, are needed for each n. The first line of (5.10) reproduces the contribution $-i/2 \sum k_l k_m (s_l + s_m)$ of the $\langle ph \rangle$ approximation, while the remaining terms of (5.10) generate all the corrections.

With (5.10) inserted under the integrals of (5.3), the "factorized" sum over n may be performed, yielding

$$
G_c(x, y|A) = i \int_0^\infty ds\, e^{-ism^2} \int \frac{d^4p}{(2\pi)^4} e^{ip\cdot(x-y)-isp^2} \cdot \int \frac{d^4K\, d^4Q}{(2\pi)^4} e^{iQ\cdot K + \frac{is}{4} K^2}
$$

$$
\cdot \int \frac{d^4z\, d^4P}{(2\pi)^4} e^{iP\cdot z} \cdot \Pi'_N \int \frac{d^4 C_N\, d^4 P_N}{(2\pi)^4} e^{i[P_N \cdot C_N - 2\frac{s}{\pi^2}\frac{C_N^2}{N^2}]}
$$

$$
\cdot \int \frac{d^4 S_N\, d^4 Q_N}{(2\pi)^4} e^{i[Q_N \cdot S_N - 2\frac{s}{\pi^2} S_N^2/N^2]}
$$

$$
\cdot \exp\left[-ig \int_0^s ds'\, A\left(y - z - Q + s'(2p - P) \right.\right.
$$

$$
\left.\left. - \sum_N{}' \left[P_N \cos\left(\frac{N\pi s'}{s}\right) + Q_N \sin\left(\frac{N\pi s'}{s}\right) \right] \right) \right].
$$

$$(5.11)$$

Now the C_N and S_N integrations may be performed immediately, and yield, for each N,

$$\left(-\frac{i\pi^6 N^4}{4s^2}\right)\exp\left[i\frac{\pi^2 N^2}{8s}(P_N^2 + Q_N^2)\right],$$

which suggests that a rescaling of all eight components is appropriate:

$$P_N \to \left(\frac{2\sqrt{s}}{N\pi}\right)P_N, \qquad Q_N \to \left(\frac{2\sqrt{s}}{N\pi}\right)Q_N,$$

leading to

$$G_c(x, y|A) = i\int_0^\infty ds\, e^{-ism^2}\int \frac{d^4p}{(2\pi)^4}\, e^{ip\cdot(x-y)-isp^2}\int \frac{d^4K\, d^4Q}{(2\pi)^4}\, e^{iQ\cdot K+\frac{is}{4}K^2}$$

$$\cdot \int \frac{d^4z\, d^4P}{(2\pi)^4}\, e^{iP\cdot z}\cdot \Pi'_N(-i)^2\int \frac{d^4P_N\, d^4Q_N}{(2\pi)^4}\cdot e^{\frac{i}{2}[P_N^2+Q_N^2]}$$

$$\cdot \exp\left[-ig\int_0^s ds'A\left(y - z - Q + s'(2p - P) - \frac{2\sqrt{s}}{\pi}\sum_N{}'\frac{1}{N}\right.\right.$$

$$\left.\left.\cdot\left[P_N\cos\left(\frac{N\pi s'}{s}\right) + Q_N\sin\left(\frac{N\pi s'}{s}\right)\right]\right)\right]. \tag{5.12}$$

Note that if any of the P_N, Q_N dependence inside A is dropped, the normalization of those integrals gives exactly the factor $+1$.

As in the reference of Note 1, we now simplify by first reflecting $z \to -z$, $P \to -P$, and then by translating: $z \to z - y$, so that all y-dependence disappears from the argument of A. Finally, the translation $z \to z + Q$ removes all Q-dependence from the argument of A, so that inegration over K and Q may be readily performed. The result is

$$G_c(x, y|A) = i\int_0^\infty ds\, e^{-ism^2}\int \frac{d^4p}{(2\pi)^4}\, e^{ip\cdot(x-y)-isp^2}\int \frac{d^4z\, d^4P}{(2\pi)^4}\, e^{iP\cdot(z-y)+is\,P^2/4}$$

$$\cdot \Pi'_N(-i)^2\int \frac{d^4P_N\, d^4Q_N}{(2\pi)^4}\, e^{\frac{i}{2}[P_N^2+Q_N^2]}$$

$$\cdot \exp\left[-ig\int_0^s ds'A\left(z + s'(2p + P') - \frac{2\sqrt{s}}{\pi}\sum_N{}'\frac{1}{N}\right.\right.$$

$$\left.\left.\cdot\left[P_N\cos\left(\frac{N\pi s'}{s}\right) + Q_N\sin\left(\frac{N\pi s'}{s}\right)\right]\right)\right], \tag{5.13}$$

and, with $q = p - p'$, it will now be convenient to pass to the momentum-space representation

$$\langle p|G_c[A]|p'\rangle = i \int_0^\infty ds\, e^{-is(m^2+p^2)} \int \frac{d^4z}{(2\pi)^4} e^{-iz\cdot q + i\frac{s}{4}q^2}$$

$$\cdot \Pi'_N(-i)^2 \int \frac{d^4 P_N\, d^4 Q_N}{(2\pi)^4} e^{\frac{i}{2}[P_N^2 + Q_N^2]}$$

$$\cdot \exp\left[-ig\int_0^s ds' A\left(z + s'(p+p') - \frac{2\sqrt{s}}{\pi}\sum_N' \frac{1}{N}\right.\right.$$

$$\left.\left.\cdot\left[P_N \cos\left(\frac{N\pi s'}{s}\right) + Q_N \sin\left(\frac{N\pi s'}{s}\right)\right]\right)\right]. \qquad (5.14)$$

Equation (5.14) differs from the $\langle ph\rangle$ approximation in two ways: (i) it contains the phase factor $\exp[isq^2/4]$; and (ii) it requires integration over all the P_N, Q_N variables. Note that the phase factor of (i) can be written as $\Pi'_N \exp[2isq^2/\pi^2 N^2]$, with each N-dependent factor inserted under its own phase-space integral.

Equation (5.14) is exact, but it is certainly not the only form possible. For example, consider the variable change $z \to z' = z + \sum_N' \alpha_N P_N$, where the α_N are a set of coefficients to be chosen such that the phase factor of (i) is removed. That is, $-iz \cdot q \to -iz \cdot q - i\sum_N' \alpha_N q \cdot P_N$, with an addition to the P_N-phase dependence which suggests a further change of variable $P_N \to P'_N = P_N - q\alpha_N$, so that

$$\frac{i}{2}P_N^2 - i\alpha_N q \cdot P_N = \frac{i}{2}P_N'^2 - \frac{i}{2}q^2\alpha_N^2. \qquad (5.15)$$

With the choice $\alpha_N = 2\sqrt{s}/\pi N$, and the property: $\sum_N' N^{-2} = \pi^2/8$, the phase factor of (i) is removed; but the new z' and old P_N dependence inside A must be re-expressed in terms of P'_N, so that the new argument of A becomes

$$z + s'(p + p') + \sum_N' \alpha_N(P'_N + q\alpha_N)$$

$$-\frac{2\sqrt{s}}{\pi}\sum_N' \frac{1}{N}\left[(P'_N + q\alpha_N)\cos\left(\frac{N\pi s'}{s}\right) + Q_N \sin\left(\frac{N\pi s'}{s}\right)\right].$$

With the aid of (5.7), the q-dependence arranges itself into

$$\left(\frac{2\sqrt{s}}{\pi}\right)^2 q \sum_N' \frac{1}{N^2}\left[1 - \cos\left(\frac{N\pi s'}{s}\right)\right] = sq\left(\frac{s'}{s}\right) = s'(p - p'),$$

so that all explicit p' dependence vanishes, leaving

$$
\langle p|G_c[A]|p'\rangle = i \int_0^\infty ds\, e^{-is(m^2+p^2)} \int \frac{d^4 z}{(2\pi)^4} e^{-iz\cdot q}
$$
$$
\cdot \Pi_N'(-i)^2 \int \frac{d^4 P_N\, d^4 Q_N}{(2\pi)^4} \cdot e^{\frac{i}{2}[P_N^2 + Q_N^2]}
$$
$$
\cdot \exp\left[-ig \int_0^s ds'\, A\left(z + 2s'p - \frac{2\sqrt{s}}{\pi} \sum_N' \frac{1}{N} \right.\right.
$$
$$
\left.\left. \cdot \left[P_N\left(\cos\left(\frac{N\pi s'}{s} \right) - 1 \right) + Q_N \sin\left(\frac{N\pi s'}{s} \right) \right] \right) \right], \quad (5.16)
$$

where the prime of P_N' has been suppressed. Clearly, (5.16) is the appropriate, exact generalization of the BN functional, (5.1).

The fact that one finds equivalent relations of the form (5.14) and (5.16) is really no surprise, for the original Fradkin representation involved the operation

$$
e^{i\int_0^s ds'\frac{\delta^2}{\delta v^2}} \cdot e^{i\int_0^s ds'v(s')\cdot p} \cdot e^{-ig\int_0^s ds'A(y-\int_0^{s'} v)}|_{v\to 0}, \quad (5.17)
$$

where p denotes the momentum used to provide a Fourier representation of $\delta(x - y + \int_0^s v)$. Now, using the relation quoted just before (4.34), (5.17) can be re-written as

$$
e^{-isp^2} \cdot e^{i\int_0^s ds'\frac{\delta^2}{\delta v^2}} \cdot e^{-ig\int_0^s ds'A(y+2s'p-\int_0^{s'} v)}|_{v\to 0}, \quad (5.18)
$$

while, from Chapter 2, the linkage operation of (5.18) can be cast into that of functional integration, in the form

$$
\mathcal{N} \int d[v]e^{\frac{i}{4}\int_0^s ds'v^2(s')} \cdot e^{-ig\int_0^s ds'A(y+2s'p-\int_0^{s'} v)}, \quad (5.19)
$$

where $\mathcal{N}$ is a normalization constant: $\mathcal{N}^{-1} = \int d[v]e^{\frac{i}{4}\int_0^s ds'v^2}$.

Imagine now expanding the 4-vector $v(s')$ in a Fourier series over the range $0 \le s' \le s$: $v(s') = -\frac{2}{\sqrt{s}}\sum_N'[P_N \sin(\frac{N\pi s'}{s}) + Q_N \cos(\frac{N\pi s'}{s})]$, so that the functional integral of (5.19) is converted into an infinite number of properly-normalized integrals over all the P_N, Q_N. The Gaussian weighting of (5.19) becomes $(1/2)\{P_N^2 + Q_N^2\}$, while the integral $-\int_0^{s'} v$ inside the argument of A produces exactly the quantity appearing in (5.16).

Both (5.14) and (5.16) are exact variants of the Fradkin representation, but it should be noted that partial summations over all N have been used in demonstrating their equivalence; approximations to (5.14) and (5.16) which involve only a finite number of N-values will not necessarily generate the same results. Because the 'zeroth' approximation to (5.14) – the neglect of all N-terms inside A – leads to $G_c^{\langle ph \rangle}[A]$, which quantity satisfies TRI (as do all of its subsequent corrections), one is moved to concentrate on approximations to (5.14). Of course, in the limit of small-momentum transfers, $|q| \ll |p|$ or $|p'|$, both forms lead to equivalent results. But, at least in potential scattering at large momentum transfers, it is known[4] that a very good approximation obtains when there is but one hard- plus many soft-interactions with the scattering potential; and here, again, $G_c^{\langle ph \rangle}[A]$ displays the correct form if only one of its Fourier momenta, k, is allowed to be large. For these reasons, we shall consider only approximations to the $G_c[A]$ of (5.14).

5.2 Finite-quadrature approximations

We now argue that the use of only a few, lowest N-values in (5.14) will lead to a respectable approximation for $G_c[A]$, and demonstrate that this claim is true in one special case where the exact, non-trivial solution is easily calculable. The motivation for this claim obtains by re-writing (5.7) in the form

$$|x| = \frac{8}{\pi^2} \sum_N' \frac{1}{N^2} \sin^2 \left(\frac{N\pi x}{2} \right), \tag{5.20}$$

and by comparing both sides of (5.20), as in Fig. 5.1. The sum over all odd N is necessary to reproduce $|x|$ exactly; but the use of just $N = 1$, or of $N = 1, 3$, etc., gives a reasonably accurate approximation to $|x|$. The error induced in $\langle p|G_c[A]|p' \rangle$ by such approximations will depend upon the relevant kinematical domains of q^2 and $(p + p')^2$, and on the way those domains depend upon the difference variables which form $|x|$. On the other hand, because this is the *only* approximation contemplated, one can certainly anticipate the order-of-magnitude of the errors which will be generated.

A glance at Fig. 5.1 makes this clear. Suppose one chooses, as the measure of error, the relative deviation of the area under each curve, between

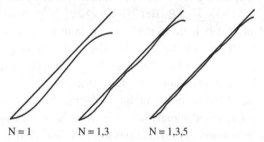

$$N = 1 \qquad\qquad N = 1,3 \qquad\qquad N = 1,3,5$$

Fig. 5.1 A plot of $(8/\pi^2) \sum_N' N^{-2} \sin^2(N\pi x/2)$ vs x, for $N = 1$, $N = 1$ and 3, $N = 1$, 3 and 5; and a comparison with x, the value of this sum over all odd N.

$x = 0$ and $x = 1$. The relative error involved in using $N = 1$ only will then be $\Delta = [\frac{1}{2} - 8\pi^2 \int_0^1 dx \sin^2(\frac{\pi x}{2})]/(\frac{1}{2}) \simeq 0.19$. When contributions from $N = 1$ and $N = 3$ are used, Δ is reduced to $\simeq 0.099$; and when $N = 1, 3,$ and 5 terms are included, Δ drops to $\simeq 0.067$. Of course, there is no guarantee that the error in any kinematical domain is not larger than these estimates; but one would expect the overall error – measured in some average way – to be reduced from the 20% of $N = 1$ to the 10% of $N = 1$ and 3, and to the 7% of $N = 1, 3, 5$. In the example below, where we have some control over the precise kinematical regions, the errors are in fact considerably smaller, suggesting that these estimates are probably upper bounds. It should perhaps again be emphasized that the only approximation contemplated in this Green's-function construction is the replacement of $|x|$ by one or another of the curves of Fig. 5.1.

Of course, these curves say nothing about the quality of the $\langle ph \rangle$ approximation, which is the "zeroth-order" approximation of this approach; the above estimates concern only the results generated by different choices of N. If *no* value of N is used, if the $-|s_l - s_m|/2$ term of $h(s_l, s_m)$ is neglected altogether, one must turn elsewhere for an estimate of that error. Fortunately, as described in the previous section, we know that $G_c^{\langle ph \rangle}[A]$ provides a very good description of small-momentum-transfer Physics, as well as a respectable description of large momentum transfers;[4] but, unfortunately, we have at present no estimate of its quality at intermediate momentum transfers.

In order to have a non-trivial, exact result with which to compare finite-N-quadrature approximations, one may turn to that soluble example mentioned in Chapter 3, wherein $A(z)$ is quadratic in its argument, which we take in the form of an inverted SHO potential, $gA(z) = gA_0 z^2 \Rightarrow -(\lambda/4)z^2$, with $\lambda > 0$. Thus, the A-dependence of (3.14) can be written as

$$-ig \int_0^s ds' A\left(y - \int_0^{s'} v\right) = i\frac{\lambda}{4}\left\{sy^2 - 2\int_0^s ds'(s - s')y_\mu v_\mu(s') + \int_0^s ds_1\right.$$

$$\left. \cdot \int_0^s ds_2 v_\mu(s_1)h(s - s_1, s - s_2)v_\mu(s_2)\right\},$$

where the linear v-dependence has been recast by Abel's transformation, and the function $h(s_1, s_2)$ is the same quantity – the smaller of the values of s_1 and s_2 – as defined in (5.5).

The functional operation of (3.14) may be performed using (2.14), and yields

$$\exp\left[-\frac{1}{2}\text{Tr} \ln(1 + \lambda K) - i\int f(1 + \lambda K)^{-1}f\right], \tag{5.21}$$

where $K_{\mu\nu}(s_1, s_2) = \delta_{\mu\nu} h(s - s_1, s - s_2)$ and $f_\mu(s') = p_\mu - \frac{\lambda}{2}(s - s')y_\mu$. To evaluate (5.21), it is convenient to introduce the Fourier series representation

$$h(s_1, s_2) = \frac{8s}{\pi^2} \sum_N' \frac{1}{N^2} \sin\left(\frac{N\pi s_1}{2s}\right) \sin\left(\frac{N\pi s_2}{2s}\right), \tag{5.22}$$

where, again, $\sum_N'$ indicates a sum over all positive, odd integers N. Using the orthogonality relation,

$$\int_0^s ds' \cos\left(\frac{N\pi s'}{2s}\right) \cos\left(\frac{M\pi s'}{2s}\right) = \frac{s}{2}\delta_{N,M},$$

one easily obtains

$$\langle s_1 | [(1 + \lambda K)^{-1}]_{\mu\nu} | s_2 \rangle$$
$$= \delta_{\mu\nu}\left[\delta(s_1 - s_2) - \frac{8\lambda s}{\pi^2} \sum_N' \left(\frac{1}{N^2 + \lambda\kappa^2}\right) \cos\left(\frac{N\pi s_1}{2s}\right) \cos\left(\frac{N\pi s_2}{2s}\right)\right],$$

where $\kappa = 2s/\pi$. Since $\sum_{\mu\nu} \delta_{\mu\nu}^2 = 4$, the first term of the exponential of (5.22) generates

$$-\frac{1}{2} \sum_{\mu\nu} \int_0^\lambda d\lambda' \int_0^s ds_1 \int_0^s ds_1 \langle s_1 | K_{\mu\nu} | s_2 \rangle \langle s_2 | [(1 + \lambda' K)^{-1}]_{\nu\mu} | s_1 \rangle$$
$$= -\frac{1}{4} \cdot 4 \cdot \sum_N' \ln(1 + \lambda\kappa^2/N^2),$$

and its exponential yields

$$[\Pi_N'(1 + \lambda\kappa^2/N^2)]^{-2} = [\cosh(s\sqrt{\lambda})]^{-2}. \tag{5.23}$$

The second exponential factor of (5.22) can be evaluated in like manner,

$$-i \int_0^s ds_1 \int_0^s ds_2 f_\mu(s_1)\langle s_1 | [(1 + \lambda K)^{-1}]_{\mu\nu} | s_2 \rangle f_\nu(s_2)$$
$$= -is\left[p^2 - \frac{\lambda s}{2}(p \cdot y) + \frac{s^3}{3}\left(\frac{\lambda}{2}\right)^2 y^2\right]$$
$$+ is\left\{ p^2[1 - \gamma(s\sqrt{\lambda})] - \frac{\lambda s}{2}(p \cdot y)\left[1 - \gamma(s\sqrt{\lambda})\gamma\left(\frac{s}{2}\sqrt{\lambda}\right)\right]\right.$$
$$\left. + \frac{\lambda^2 s^2 y^2}{12} - \frac{\lambda y^2}{4}[1 - \gamma(s\sqrt{\lambda})]\right\}, \tag{5.24}$$

where $\gamma(x) \equiv \frac{\tanh(x)}{x} = \frac{8}{\pi^2} \sum_N' \frac{1}{N^2 + 4x^2/\pi^2}$.

Combining all factors, one obtains

$$\langle p|G_c[A]|p'\rangle = i \int_0^\infty \frac{ds\, e^{-ism^2}}{\cosh^2(s\sqrt{\lambda})} \cdot \int d^4 y\, e^{-iq\cdot y - isp^2\gamma(s\sqrt{\lambda})}$$

$$\cdot e^{i[\frac{\lambda s^2}{2}(p\cdot y)\gamma(s\sqrt{\lambda})\gamma(\frac{s}{2}\sqrt{\lambda}) + \frac{\lambda s y^2}{4}\gamma(s\sqrt{\lambda})]}. \tag{5.25}$$

There is one remaining Gaussian integration (over y) to be performed in (5.25), straightforward but tedious; and the result is

$$\langle p|G_c[A]|p'\rangle = \left(\frac{4\pi i}{\lambda}\right)^2 \cdot \int_0^\infty ds\, [s^2 D(s\sqrt{\lambda})]^{-1} \cdot e^{-ism^2}$$

$$\cdot \exp\left\{ -\frac{i}{2\lambda}\left[P^2 \tanh\left(\frac{s}{2}\sqrt{\lambda}\right) + q^2 \coth\left(\frac{s}{2}\sqrt{\lambda}\right) \right] \right\}, \tag{5.26}$$

with $P = p + p'$, $q = p - p'$, and $D(x) = [\sinh(x)/x]^2$. Incidentally, this exhibition of $G_c[A]$ in terms of an integral over proper time is what is here meant by the phrase "exact solution".

In contrast, the $\langle ph\rangle$ approximation requires but one Gaussian integration, which yields

$$\langle p|G_c^{\langle ph\rangle}[A]|p'\rangle = \left(\frac{4\pi i}{\lambda}\right)^2 \int_0^\infty \frac{ds}{s^2} e^{-ism^2} \cdot e^{-i[s\frac{P^2}{4}(1-\frac{\lambda s^2}{12}) + \frac{q^2}{\lambda s}(1+\frac{\lambda s^2}{4})]}, \tag{5.27}$$

while the $\langle ph\rangle + (N = 1) = \langle ph|1\rangle$ contributions require an extra pair of Gaussian integrations, over P_1 and Q_1, and generate

$$\langle p|G_c^{\langle ph|1\rangle}[A]|p'\rangle = \left(\frac{4\pi i}{\lambda}\right)^2 \int_0^\infty \frac{ds\, e^{-ism^2}}{s^2 D_1(s\sqrt{\lambda})} e^{-i[s\frac{P^2}{4}A(s\sqrt{\lambda}) + \frac{q^2}{\lambda s}B(s\sqrt{\lambda})]}, \tag{5.28}$$

where

$$D_1(s\sqrt{\lambda}) = \left[1 + \frac{\lambda s^2}{\pi^2}\left(1 - \frac{8}{\pi^2}\right)\right]^2 \cdot \left[1 + \frac{\lambda s^2}{\pi^2}\right]^2,$$

$$A(s\sqrt{\lambda}) = 1 - \frac{\lambda s^2}{12} + \frac{8}{\pi^2}\left(\frac{\lambda s^2}{\pi^2}\right)^2 \left[1 + \frac{\lambda s^2}{\pi^2}\right]^{-1},$$

and

$$B(s\sqrt{\lambda}) = 1 + \frac{8\lambda s^2}{\pi^4}\left[1 + \frac{\lambda s^2}{\pi^2}\left(1 - \frac{8}{\pi^2}\right)\right]^{-1}.$$

We now compare the three results, (5.26), (5.27), and (5.28). In general, as q^2 decreases, the dominant contributions of each must come from small s – that is, small $s\sqrt{\lambda}$ – in which region all three integrands exhibit the same s-dependence, defined in part by the limits

$$\frac{P^2}{2\sqrt{\lambda}} \tanh\left(\frac{s\sqrt{\lambda}}{2}\right) \simeq \frac{s}{4}P^2\left[1 - \frac{\lambda s^2}{12} + \cdots\right];$$

$$\frac{q^2}{2\lambda} \coth\left(\frac{s\sqrt{\lambda}}{2}\right) \simeq \frac{q^2}{s\lambda}\left[1 + \frac{\lambda s^2}{12} + \cdots\right].$$

This is no surprise, because we expect such convergence in the limit of small momentum transfer. In each integrand, the dominant contributions will come from the range $s_{max} > s > s_{min}$; and for each, it is clear that $s_{min} \sim q^2/\lambda$.

Let us now suppose that $m = 0$, to insure that the s_{max} of each integral depends upon its own distinctive features; the argument to follow will only be enhanced if $m \neq 0$. The integral of the exact solution is cut off by the D-factor when $s \sim s_{max} \sim (\lambda)^{-1/2}$, whereas in (5.27) and (5.28) it is controlled (assuming $|P^2| > q^2$) by P^2: $s_{max} \sim 4/|P^2|$. (In the region near $\lambda s^2 \sim 12$, there can be a large contribution to the $\langle ph \rangle$ integrand, although the latter's denominator factor of s^{-2} means that such a contribution is weighted with a factor of $\lambda/12$, which is small if $\lambda < 1$. If $q^2 \sim |P^2|$, however, the $(-isq^2/4)$ phase factor will serve to define $s_{max} \sim 4/q^2 \sim 4/|P^2|$.)

In order to have the same $s_{max} \sim (\lambda)^{-1/2}$ for each integral, let us choose the simple (and physically reasonable) kinematical restriction: $|P^2|/4 \sim (\lambda)^{1/2} > q^2$. In the contributing region, between $s_{min} \sim q^2/\lambda$ and $s_{max} \sim (\lambda)^{-1/2}$, all of the exponential factors are essentially the same, and the only real distinction between these three results can arise from the different denominator factors, D, D_0, and D_1. We evaluate these in a crude way by calculating the first term in each small $-\lambda s^2$ expansion, assigning to it its largest value at $s = s_{max}$: $D_0 \equiv 1$, $D(s\sqrt{\lambda}) \sim 1 + \lambda\frac{s^2}{3} + \cdots \rightarrow 4/3$,

$$D_1(s\sqrt{\lambda}) \sim 1 + 2\frac{\lambda s^2}{\pi^2}(2 - 8/\pi^2) + \cdots \rightarrow 1.24.$$

The maximum relative error of the $\langle ph \rangle$ approximation is then

$$\Delta_0 = \left(\frac{4}{3} - 1\right)\Big/\left(\frac{4}{3}\right) = \frac{1}{4} \rightarrow 25\%,$$

but the maximum relative error of the $\langle \mathrm{ph}|1\rangle$ result will be

$$\Delta_1 = \left(\frac{4}{3} - 1.24\right) \Big/ \left(\frac{4}{3}\right) \simeq 0.068 \rightarrow 7\%.$$

In this crude way, and for this special choice of potential and kinematical restriction, one sees that the $\langle \mathrm{ph}|1\rangle$ approximation will be an improvement of the $\langle \mathrm{ph}\rangle$ result, and that the latter is a decent representation of the exact solution to within an overall error of 25%.

5.3 Exact and approximate vectorial interactions

Our starting point is again the exact Fradkin representation for a causal QED fermion propagator moving in an arbitrary external (or "background") field specified by the vector potential $A_\mu(z)$, as expressed in (3.32).[3] Inserting a Fourier representation for the function of (3.31), the linkage operator $\exp[i \int_0^s ds' \delta^2/\delta v^2]$ is required to link in a pairwise fashion, that is, to "factor pair" all the v_μ upon which it acts. The essential difference between the present vector and previous scalar case, given in terms of a scalar potential $A(y - \int_0^{s'} v)$, is the appearance of the vector forms $v_\mu(s')A_\mu(y - \int_0^{s'} v)$, and the necessity of factor-pairing *all* pairs of the v-dependence; in particular, one must retain those linkages of $v_\mu(s')$ and $A_\nu(y - \int_0^{s''} v)$, which will appear upon expansion in powers of g. That is, because the linkage operator can only link v-dependence at the same value of s', a factor pairing between $v_\mu(s')$ and its immediate coefficient $A_\mu(y - \int_0^{s'} v)$ must vanish. But in the linkages of $\int_0^s ds_1 v_\mu(s_1)A_\mu(y - \int_0^{s_1} v) \cdot \int_0^s ds_2 v_\nu(s_2)A_\nu(y - \int_0^{s_2} v)$ there will be contributions coming from pairings of $v_\mu(s_1)$ and $A_\nu(y - \int_0^{s_2} v)$, and from $A_\mu(y - \int_0^{s_1} v)$ and $v_\nu(s_2)$, depending on the relative size of s_1 and s_2. Of course, there are also the relatively simple pairings between the $v_\mu(s_1)$ and the $v_\nu(s_2)$, as between $A_\mu(y - \int_0^{s_1} v)$ and $A_\nu(y - \int_0^{s_2} v)$, which were fully analyzed in Section 5.1; but a novel structure now appears when we include pairings between the $v_\mu(s')$ and the $A_\nu(y - \int_0^{s''} v)$.

To see how this goes, we trivially replace the exponential factor of (3.31),

$$\exp\left[-ig \int_0^s ds' v_\mu(s')A_\mu\left(y - \int_0^{s'} v\right)\right],$$

by

$$\int d[\phi]\delta[\phi - v] \exp\left[-ig \int_0^s ds' \phi_\mu(s')A_\mu\left(y - \int_0^{s'} v\right)\right],$$

and insert a by-now standard representation of the δ-functional,

$$N' \int d[\Omega] \exp\left[i \int_0^s ds' \Omega_\mu(s')[\phi_\mu(s') - v_\mu(s')]\right],$$

with N' an appropriate normalization constant. In this way, the "external" $v_\mu(s')$ terms may be isolated and removed from their immediate neighbors $A_\mu(y - \int_0^{s'} v)$, so that $G_c[A]$ may be re-written as

$$G_c(x, y|A) = i \int_0^\infty ds\, e^{-ism^2} \int \frac{d^4 p}{(2\pi)^4} e^{ip\cdot(x-y)} \left[m - \gamma \cdot \frac{\delta}{\delta v(s)}\right] \cdot e^{i\int_0^s ds'\,\delta^2/\delta v^2}$$

$$\cdot \left(\exp\left[g \int_0^s ds'\sigma \cdot F\left(y - \int_0^{s'} v\right)\right]\right)_+$$

$$\cdot N' \int d[\Omega] \int d[\phi] e^{i\int_0^s ds'\,\Omega_\mu(s')\phi_\mu(s')}$$

$$\cdot e^{i\int_0^s v(s')\cdot[p-\Omega(s')]} \cdot e^{ig\int_0^s ds'\,\phi_\mu(s')A_\mu(y-\int_0^{s'} v)}\big|_{v\to 0}. \qquad (5.29)$$

How does the $G_c[A]$ of (5.29) differ from the scalar $G_c[A]$ of (3.14)? Aside from the vectorial and spinorial indices, both Green's functions have the same formal dependence upon the Fradkin v_μ, except that the former requires the additional $N' \int d[\phi]d[\Omega]$ integrals, and that its exponential coefficient of v_μ is $p_\mu - \Omega_\mu$, rather than just p_μ. Imagine performing all the operations of Section 5.1 at this point; one would obtain

$$\langle p|G_c[A]|p'\rangle = i \int_0^\infty ds\, e^{-ism^2} N' \int d[\Omega]\, d[\phi] e^{-i\int_0^s ds'\,[p-\Omega(s')]^2}$$

$$\cdot \int \frac{d^4 z}{(2\pi)^4} e^{-iz\cdot q + i\frac{sq^2}{4}} \Pi_N'(-i)^2 \int \frac{d^4 P_N\, d^4 Q_N}{(2\pi)^4} e^{\frac{i}{2}[P_N^2 + Q_N^2]}$$

$$\cdot e^{i\int_0^s ds'\,\phi_\mu(s')[\Omega_\mu(s') - gA_\mu(\zeta(s') - 2\int_0^{s'} ds''\,\Omega(s''))]}$$

$$\cdot [m - i\gamma \cdot (p - \Omega(s))]\left(\exp\left[g \int_0^s ds'\sigma\right.\right.$$

$$\left.\left.\cdot F\left(\zeta(s') - 2\int_0^{s'} ds''\,\Omega(s'')\right)\right]\right)_+, \qquad (5.30)$$

where $\zeta(s') = z + s'(p + p') - \frac{2\sqrt{s}}{\pi} \sum_N' \frac{1}{N}[P_N \cos(\frac{N\pi s'}{s}) + Q_N \sin(\frac{N\pi s'}{s})]$; this is simply a transcription of (5.14), including the extra functional integrals over ϕ and Ω, and the replacement of p by $p - \Omega(s')$. Now, however, the $\int d[\phi]$ is immediate, and generates the δ-functional

$$\delta\left[\Omega(s') - gA\left(\zeta(s') - 2\int_0^{s'} ds''\Omega(s'')\right)\right]$$

which restricts admissible $\Omega(s')$ to those functions which satisfy the nonlinear equation, or "map",

$$\Omega_\mu(s') = gA_\mu\left(\zeta(s') - 2\int_0^{s'} ds''\Omega(s'')\right), \qquad (5.31)$$

and it is this requirement, as discussed in the next chapter, which opens the possibility of chaos, or of chaotic effects, appearing in these otherwise well-behaved Green's functions of potential theory.

Integration $\int d[\Omega]$ over this δ-functional is performed by changing variables to an $f(s')$ defined as the argument of the δ-functional of (5.30), and yields

$$[\det(\delta f/\delta\Omega)]^{-1} = \exp[-\mathrm{Tr}\,\ln(\delta f/\delta\Omega)], \qquad (5.32)$$

evaluated at that Ω which is a solution of (5.31). If there is more than one such $\Omega(s')$ satisfying (5.31), a summation must be made over all such solutions. With

$$\frac{\delta f_\mu(s')}{\delta\Omega_\nu(s'')} = \delta_{\mu\nu}\delta(s' - s'') + 2g\theta(s' - s'')\frac{\partial}{\partial z_\nu}A_\mu\left(\zeta(s') - 2\int_0^{s'} ds''\Omega(s'')\right),$$

and because of the "retardedness" of the $\theta(s' - s'')$ factors, the $\mathrm{tr}\,\ln(\delta f/\delta\Omega)$ of (5.31) may be replaced by its lowest-order term (which vanishes in the Lorentz gauge), so that

$$\langle p|G_c[A]|p'\rangle$$

$$= i\int_0^\infty ds\,e^{-ism^2}\int\frac{d^4z}{(2\pi)^4}e^{-iz\cdot q + isq^2/4}\Pi'_N(-i)^2$$

$$\cdot\int\frac{d^4P_N\,d^4Q_N}{(2\pi)^4}e^{\frac{1}{2}[P_N^2+Q_N^2]}\cdot e^{-i\int_0^s ds'[p-\Omega(s')]^2}$$

$$\cdot\exp\left[-g\int_0^s ds'\frac{\partial}{\partial z_\mu}A_\mu\left(\zeta(s') - 2\int_0^{s'}\Omega\right)\right]\cdot\{m - i\gamma\cdot[p - \Omega(s)]\}$$

$$\cdot\left(\exp\left[g\int_0^s ds'\sigma\cdot F\left(\zeta(s') - 2\int_0^{s'}\Omega\right)\right]\right)_+. \qquad (5.33)$$

Not only is the $G_c[A]$ of (5.33) more complicated than that of the corresponding scalar Green's function, but it contains something quite new: the map of (5.31), whose solution must be obtained as a functional of z, s, P_N, and Q_N, and then substituted into (5.33). In the paper of Note 5 it is emphasized that this double complexity, while gauge dependent, is not a gauge-dependent artifact. However, the most striking aspect of this potential-theory representation is the possibility of chaos appearing in solutions of the map (that is, the appearance of ultra-sensitive dependence of $\Omega(s)$ on initial conditions, which are given in terms of z, and the P_N, Q_N), a behavior which could wreak havoc in any numerical integration of $G_c[A]$ (given in terms of integrations over these quantities) and in the correlation functions of QFT constructed from such $G_c[A]$. For those who believe in QFT, this is a horrifying prospect; but, happily, one which is apparently removed upon calculating the totality of radiative corrections (that is, quantum fluctuations of the A_μ fields) which define such correlation functions. How this comes about, and what it may portend for classical as well as quantum chaos, is briefly described in the next chapter.

5.4 The Stojkov variation

A completely independent variation of the Fradkin representation has been proposed by Stojkov,[6] one which does not employ the expansion (in g) and re-summation of Section 5.1, and which gives results comparable to, and frequently simpler than, those of the preceding sections. We give a very brief sketch of the basic method, and refer the interested reader to the original paper for further approximations and applications.

The essential idea is to substitute the expansion

$$v_\mu(s') = V_\mu + \sum_{n=1}^{\infty} \left[Q_\mu^{(n)} \cos(n\omega s') + P_\mu^{(n)} \sin(n\omega s') \right] \qquad (5.34)$$

into Fradkin's exact (scalar) representation (3.14), where $\omega = \frac{2\pi}{s}$, and V_μ is a constant 4-vector. One requires the (ad hoc) constraint $\int_0^s ds' v_\mu(s') = s V_\mu$, leaving the coefficients of the oscillatory modes unconstrained. Properly normalized basis functions on the interval 0–s are given by

$$\phi_n(s') = \begin{cases} \sqrt{\frac{1}{s}}, & n = 0 \\ \sqrt{\frac{2}{s}}\cos(n\omega s'), \quad \sqrt{\frac{2}{s}}\sin(n\omega s'), & n \geq 1, \end{cases}$$

so that the corresponding completeness statement is given by

$$\delta(s_1 - s_2) = \sum_\lambda \phi_\lambda(s_1)\phi_\lambda(s_2) = \frac{1}{s} + \frac{2}{s}\sum_{n=1}^\infty \cos(n\omega[s_1 - s_2]). \quad (5.35)$$

Then, requiring that $\frac{\delta v_\mu(s_1)}{\delta v_\nu(s_2)} = \delta_{\mu\nu}\delta(s_1 - s_2)$, one obtains a decomposition of the functional derivative,

$$\frac{\delta}{\delta v_\mu(s_1)} = \frac{1}{s}\frac{\partial}{\partial V_\mu} + \frac{2}{s}\sum_{n=1}^\infty \left\{ \cos(n\omega s_1)\frac{\partial}{\partial Q_\mu^{(n)}} + \sin(n\omega s_1)\frac{\partial}{\partial P_\mu^{(n)}} \right\}, \quad (5.36)$$

and one can then calculate

$$\int_0^s ds' \frac{\delta^2}{\delta v(s')^2} = \frac{1}{s}\frac{\partial^2}{\partial V^2} + \frac{2}{s}\sum_{n=1}^\infty \left\{ \frac{\partial^2}{\partial Q_n^2} + \frac{\partial^2}{\partial P_n^2} \right\},$$

$$\int_0^t ds' v_\mu(s') = V_\mu t + \frac{s}{2\pi}\sum_{n=1}^\infty \frac{1}{n}\left\{ Q_\mu^{(n)}\sin(n\omega t) + P_\mu^{(n)}[1 - \cos(n\omega t)] \right\},$$

and

$$\int_0^s ds' v_\mu(s') = sV_\mu.$$

It then follows that

$$G_c(x, y|A) = \frac{1}{16\pi^2}\int_0^\infty ds\, e^{-ism^2 + iz^2/4s} \cdot e^{\frac{2i}{s}\sum_{n=1}^\infty \{\frac{\partial^2}{\partial Q_n^2} + \frac{\partial^2}{\partial P_n^2}\}}$$

$$\cdot \exp\left[-ig\int_0^s ds' A(\xi(s'/s)) + \frac{s}{2\pi}\sum_n \frac{1}{n}\{ Q^{(n)}\sin(n\omega s') \right.$$

$$\left. + P_\mu^{(n)}[1 - \cos(n\omega s')]\} \right], \quad (5.37)$$

where $\xi_\mu(\lambda) = x_\mu\lambda + (1 - \lambda)y_\mu$. Equation (5.37) is a convenient variant of the Fradkin representation if one intends to approximate $G_c[A]$ by a sequence of terms corresponding to corrections to the straight-line path ξ between x and y; that lowest order approximation in which all the P_n, Q_n are neglected is called $G_c^{(0)}[A]$ by Stojkov, and is somewhat simpler to work with in configuration space than is the corresponding $G_c^{(\mathrm{ph})}[A]$, although the latter appears to be more convenient in momentum space. Stojkov discusses correction to $G_c^{(0)}[A]$, as well as similar approximations to the $L[A]$ of this problem; and he provides a most useful section in which the exact Fradkin representation is recast in the form of a path integral. In this paper, an example of the "triviality" of the scalar interaction $\mathcal{L}' = -gA^4$ is also given in the context of the $\langle 0 \rangle$ approximation.

Notes

1 H. M. Fried and Y. Gabellini, *Phys. Rev. D* **51** (1995) 890 and 906.
2 HMF#1.
3 HMF#2.
4 L. I. Schiff, *Phys. Rev.* **103** (1956) 443, and **104** (1956) 1481.
5 H. M. Fried, Y. Gabellini, and B. H. J. McKellar, *Phys. Rev. D* **51** (1995) 7083, and *Phys. Rev. Letters* **74** (1995) 4373.
6 P. Stojkov, in *New Approximations to the Fradkin Representation for Green's Functions*, hep-th/9706135, 18 June 1997. While a second-year graduate student in Theoretical Physics at Brown, Stojkov produced his solution a few days after learning about the results of Fried and Gabellini.[1]

6

Quantum chaos and vectorial interactions

The possibility of quantum chaos for vectorial interactions is here described in some detail, along with its apparent suppression when the radiative corrections of quantum field theory (quantum fluctuations of the classical, "external" electromagnetic field) are introduced. Based on these quantum-mechanical ideas, an application is made to classical-chaotic systems, of perhaps the simplest form – a forced Duffing model, without damping – where it is found that the chaos is first suppressed and then (apparently) removed by introducing couplings to random and/or chaotic sources. This may be characterized as "quantum mechanics with $\hbar \sim 1$", and suggests a brute-force method by which the chaos of a classical system may be at least diminished. A similar effect is noted for a different classical system that displays chaos – a pair of coupled oscillators – independently of any external forcing.

6.1 First-quantization chaos

The reader is now asked to return to the $G_c[A]$ representations for vectorial interactions of the previous chapter; for simplicity, the arguments of this chapter are presented only for QED in a relativistic context, but have obvious generalizations to non-relativistic QED, and to relativistic QCD.

The map (5.31) defines the quantity $\Omega_\mu(s')$, which is needed for the explicit solution of (5.33). It is the existence of such a map which carries with it the inescapable possibility of chaotic behavior, at least in the present context of vectorial interactions in potential theory. The analysis used here is given directly in terms of proper time τ, of which $x_\mu(\tau)$ is a function. In principle, one might expect nonlinear quantum systems to reflect at least partially the chaotic nature of their classical limits. Indeed, solutions for classical, charged particles moving

93

in a specified Maxwell field $F_{\mu\nu}(x)$ described by

$$\frac{d^2 x_\mu}{d\tau^2} = \frac{g}{mc} \sum_\nu \frac{dx\nu}{d\tau} F_{\mu\nu}(x) \tag{6.1}$$

for appropriately nonlinear x-dependence of $F_{\mu\nu}$ must be expected to display the chaotic behavior now well-documented in a variety of fields, and by a variety of methods.[1,2] We comment below that, with one important modification, the Green's-function map (5.31) may be related to a "first integral" of the classical (6.1).

High-energy Physics has heretofore escaped the impact of classical chaotic dynamics because of its fortunate ability to rely on perturbative expansions, which destroy the non-perturbative arguments leading to the possibility of chaos. Even certain non-perturbative approximations, such as the standard eikonal methods,[3] remove the possibility of chaos because they destroy the needed, nonlinear structure of relevant maps. This can be seen immediately from (5.31), whose perturbative expansion in effect removes $\int_0^{s'} \Omega$ from the argument of A_μ,

$$\Omega_\mu(s') \approx g A_\mu(\zeta(s')) - 2g^2 \frac{\partial A}{\partial z_\mu}(\zeta(s')) \cdot \int_0^{s'} ds'' A_\nu(\zeta(s'')) + \cdots, \tag{6.2}$$

so that one sums a g^n expansion for Ω_μ, as in (6.2), rather than solving an integral or differential equation. An eikonal approximation, on the other hand, would replace the $\int_0^{s'} ds'' \Omega(s'')$ in the argument of A by $s' \bar{\Omega}_\mu$, where $\bar{\Omega}_\mu$ is an appropriately chosen, averaged 4-velocity suggested by the specific scattering problem. For both approximation schemes, the repeated "feedback" obtained from the Ω dependence within A_μ is missing, as will be any suggestion of chaotic behavior in the final results. One sees below that strict attention to the exact forms of these vector interactions, especially when using the non-perturbative, finite-N-quadrature approximations of the previous chapter, must finally bring the possibility of chaos into the realm of high-energy Physics.[4]

In a Green's-function context, this possibility appears in a most efficient way, for it is specified not in terms of time, nor of space, nor by a mixed partial-differential-equation formulation, but in terms of proper time. This suggests that chaotic behavior in proper time, when restricted to lie inside the light cone, will correspond to (temporal) chaos; while such behavior outside the light cone may refer to a form of (spatial) turbulence. (It should be noted that Schwinger–Fradkin representations involving a variable analogous to relativistic "proper time" can be written for non-relativistic problems.) We shall not elaborate further on this distinction here, but only note that such a Green's-function description in terms of proper time would seem to be an obvious method for simultaneously describing both chaos and turbulence.[5]

For more than a decade there have been examples of "quantum chaos" for a particle whose motion is in some, semi-classical sense partially governed by Bohr–Sommerfeld quantization. One of these examples,[6] the case of an electron moving in a Coulomb potential upon which is superimposed a magnetic field, will be used for illustration of the way in which chaos can appear in the present context. By the criterion of the present discussion, the "semi-classical" Green's function of this example does not appear to be chaotic; but when at least a part of the quantum fluctuations corresponding to a finite number of the P_N, Q_N terms are retained, one appears to obtain certifiably chaotic behavior.

We first ask if there is any connection between the map (5.31) and the classical equation of motion (6.1). Let us initially approximate $G_c[A]$, and hence (5.31), by the neglect of all quantum fluctuations exhibited by the P_N, Q_N terms inside the argument of A_μ and $F_{\mu\nu}$ – this is the QED version of what was termed the $\langle$ph$\rangle$ approximation of Chapter 5 – and, secondly, by imagining that the momentum transfer q is small, so that the difference between p and p' is not important. This latter step, of replacing $G_c\langle$ph$\rangle[A]$ by $G_{\mathrm{BN}}[A]$, is not essential, but makes for conceptual simplicity.

The length $\zeta(s')$ of (5.30) then becomes $z + s'(p + p') \to z + 2s'p$, and we switch to a proper-time variable with proper dimensions, by the replacement of s' by $\tau/2m$. With the representation $\Omega_\mu = \frac{\mathrm{d}X_\mu}{\mathrm{d}\tau}$, the map now reads

$$\frac{\mathrm{d}X_\mu}{\mathrm{d}\tau} = \frac{g}{2m} A_\mu\left(z + \tau\frac{p}{m} - 2\left[X\left(\frac{\tau}{2m}\right) - X(0)\right]\right),$$

but it will be more convenient to denote the argument of A_μ by x, and to write the equivalent

$$\frac{\mathrm{d}x_\mu}{\mathrm{d}\tau} = \frac{p_\mu}{m} - \frac{g}{m} A_\mu(x). \tag{6.3}$$

Then

$$\frac{\mathrm{d}^2 x_\mu}{\mathrm{d}\tau^2} = -\frac{g}{m}\sum_\nu \frac{\mathrm{d}x_\nu}{\mathrm{d}\tau}\frac{\partial}{\partial x_\nu}A_\mu(x) = \frac{g}{m}\sum_\nu \frac{\mathrm{d}x_\nu}{\mathrm{d}\tau}\left[F_{\mu\nu}(x) - \frac{\partial}{\partial x_\mu}A_\nu(x)\right], \tag{6.4}$$

and inserting (6.3) into (6.4), there follows

$$\frac{\mathrm{d}^2 x_\mu}{\mathrm{d}\tau^2} = \frac{g}{m}\sum_\nu \frac{\mathrm{d}x_\nu}{\mathrm{d}\tau}F_{\mu\nu}(x) + 2\sum_\nu \frac{\mathrm{d}x_\nu}{\mathrm{d}\tau}\frac{\partial}{\partial x_\mu}\left(\frac{\mathrm{d}x_\nu}{\mathrm{d}\tau}\right). \tag{6.5}$$

But the last RHS term of (6.5) may be rewritten as

$$\frac{\partial}{\partial x_\mu} \sum_\nu \left(\frac{dx_\nu}{d\tau}\right)^2,$$

which quantity must vanish if the 4-velocity $v_\mu = dx_\mu/d\tau$ is to represent a particle on its mass shell (or energy shell, in a non-relativistic context), an association which is certainly compatible with the remaining terms of (6.5), and which we shall assume temporarily. The result is then just (6.1), showing that, in an appropriate semi-classical limit, the map (5.31) is compatible with the standard, classical equation of motion, if only the mass-shell property of the particle were guaranteed (rather than assumed). In fact, the map might be called a "first integral" of (6.1), since it involves terms of one derivative fewer, although appearing in the decidedly non-trivial form of a map.

For reasons which will become clear immediately, we allow the magnetic field to vary in two transverse directions, by introducing a function $\phi(x_\perp^2)$ into the expression for the vector potential,

$$A_\mu(x) = \frac{B}{2}\left[x_1\delta_{\mu 2} - x_2\delta_{\mu 1}\right]\phi(x_\perp^2) + \frac{iZg}{r}\delta_{\mu 4}, \tag{6.6}$$

where $r = [\sum_{i=1}^3 x_i^2]^{1/2}$, $x_\perp^2 = x_1^2 + x_2^2$, and the 4-vector notation used here is: $a_\mu = (a, ia_0)$.

Substitution of (6.6) into the map (6.3) leads to

$$\frac{dx_\mu}{d\tau} = v_\mu - \frac{g}{m}A_\mu(x) \equiv f_\mu(x),$$

or, in component form, with $v_\mu = p_\mu/m$, $x_4 = ix_0 = it$,

$$x_\mu(\tau) = z_\mu + \tau v_\mu - 2[X_\mu(\tau) - X_\mu(0)], \qquad \omega = gB/2m,$$

$$\frac{dx_1}{d\tau} = v_1 + \omega x_2\phi(x_\perp^2) = f_1(x),$$

$$\frac{dx_2}{d\tau} = v_2 - \omega x_1\phi(x_\perp^2) = f_2(x)$$

$$\frac{dx_3}{d\tau} = v_3 = f_3, \qquad \frac{dt}{d\tau} = v_0 + \frac{Zg^2}{mr} = f_0. \tag{6.7}$$

We have neglected the quantum fluctuations specified by variations of the P_N, Q_N terms, so that $x_\mu(\tau)$ acts as the effective position and time coordinates of a particle with momentum p. Note that the 4-velocities calculated from (6.7) will not, with constant v_μ, satisfy the mass-shell condition.

To test for chaos, one is instructed[1,2] to calculate the local, or instantaneous, Lyapunov exponents λ^α as the eigenvalues of the Jacobian of continuous transformation: $\det |\frac{\delta f}{\delta x} - \lambda|$, and then to average the λ^α – there is a variety of ways to do this – over a sufficiently long τ-interval (such as a period, or periodic orbits), to be able to see if any $\langle \lambda^\alpha \rangle$ may be considered a positive constant over that interval; if so, the system is expected to be unstable, and perhaps chaotic. In this case, the calculation of the local exponents is straightforward, yielding two zero exponents and a pair which satisfy

$$\lambda = \pm i\omega\phi\left[1 + 2x_\perp^2(\phi'/\phi)\right]^{1/2}. \tag{6.8}$$

From (6.8), it is clear that a constant magnetic field, $B \to B_0$, for $\phi \to 1$, corresponds to imaginary roots, and thus to pure oscillations in the distance between neighboring trajectories. Hence, this Green's function, in its semi-classical limit, does not display the chaotic behavior found in the classical limit; and the reason for the difference appears to be that the "motion" to which (6.7) corresponds does not contain the needed mass-shell restriction. If the map (6.6) is altered so that the mass-shell condition is maintained, one finds equations reminiscent of Hamilton–Jacobi theory, in which the same analysis does lead to the possibility of chaos.

If we allow the ratio ϕ'/ϕ to become sufficiently negative to convert the square root of (6.8) to imaginary values, one might hope to gain a positive eigenvalue. However, if the magnetic field is allowed to fall away to zero, or even to change sign, the "recurrence" of the motion will be lost, as the particle moves out to larger and larger $x_\perp$ values. What is needed, then, is another augmentation so that the particle is bound in a narrow $x_\perp$ range, and where the possibility exists that the average value of one exponent will be positive.

This motion, however, appears to be integrable,[7] with explicit, oscillatory solutions possible in relevant energy ranges. In order to achieve chaotic motion in the present context, one must retain at least a part of the oscillatory τ-dependence contained in the P_N, Q_N terms. Then the problem is no longer integrable, overlapping frequencies will appear, and chaotic motion almost always occurs. Examples of this are well known in the mathematical literature of chaos,[2] and in the present context signify that there is a fundamental uncertainty built into any of these approximations to the exact Green's functions of first quantization which contain vectorial interactions. Note that the sum over *all* such P_N, Q_N terms need not lead to chaos; it is only when these corrections to the semi-classical behavior are taken into account, involving specific, oscillatory τ-dependence, that one can expect KAM tori, on which a phase point would move, to be disrupted, and Arnold diffusion of that phase point to appear, heralding the onset of chaos.

6.2 Chaos suppression in second quantization

Consider the exact representation of Chapter 5 for the calculation of the simplest, two-point function of QED, the dressed fermion propagator in an external field,

$$\langle p|S_c'[A^{\text{ext}}]|p'\rangle = e^{\mathcal{D}_A}\langle p|G_c[A^{\text{ext}} + A]|p'\rangle \cdot \left.\frac{e^{L[A^{\text{ext}}+A]}}{\langle S[A^{\text{ext}}]\rangle}\right|_{A\to 0}, \qquad (6.9)$$

where $\mathcal{D}_A = -\frac{i}{2}\int \frac{\delta}{\delta A_\mu} D_{c,\mu\nu}\frac{\delta}{\delta A_\nu}$, $D_{c,\mu\nu}$ is the (bare) photon propagator (containing a mass μ, as temporary insurance against any IR difficulty), $L[A] =$ $\text{Tr}\ln(1 + ig\gamma \cdot AS_c)$ is the fermion closed-loop functional, associated with the vacuum-to-vacuum amplitude as normalization factor $\langle S[A^{\text{ext}}]\rangle =$ $e^{\mathcal{D}_A}e^{L[A^{\text{ext}}+A]}|_{A\to 0}$. As described and used in previous chapters, the functional linkage operations here are exactly equivalent to Gaussian-weighted functional integration,

$$e^{\mathcal{D}_A}\mathcal{F}[A]|_{A\to 0} = N\int d[A]e^{-\frac{1}{2}\int A(\mu^2-\partial^2)A}\mathcal{F}[A],$$

where $\mu^2 - \partial^2 = D_c^{-1}$, and N^{-1} is the same functional integral but with $\mathcal{F}[A]$ replaced by unity.

It will be simplest to work in the "quenched" approximation, neglecting the A fluctuations of $L[A^{\text{ext}} + A]$; this means the replacement of $L[A^{\text{ext}} + A]$ by $L[A^{\text{ext}}]$, and $\langle S[A^{\text{ext}}]\rangle$ by $\exp\{l[A^{\text{ext}}]\}$ in (6.9). This simplification is not essential to subsequent arguments, for one can include arbitrary powers of $L[A^{\text{ext}} + A]$ fluctuations with unchanged conclusions. The same remark is true for all other n-point functions of the theory, as will be made clear below.

We now ask the reader to return to the exact representation (5.30) of $G_c[A]$ before the final $\int d[\phi] \int d[\Omega]$ integrations were performed; in compact notation, this can be written as

$$\langle p|G_c[A]|p'\rangle = i\int_0^\infty ds\, e^{-ism^2}\int d^4 z\, e^{-iq\cdot z+isq^2/4}\Pi_N'\frac{(-i)^2}{(2\pi)^4}$$

$$\cdot \int d^4 P_N d^4 Q_N e^{\frac{1}{2}(P_N^2+Q_N^2)}\cdot N'\int d[\phi]\int d[\Omega]\mathcal{F}[\Omega]$$

$$\cdot \exp\left\{i\int_0^s ds'\phi_\mu(s')\left[\Omega_\mu(s') - gA_\mu\left(\zeta(s') - 2\int_0^{s'}\Omega\right)\right]\right\},$$

$$(6.10)$$

where $\mathcal{F}[\Omega]$ represents all the remaining Ω-dependence visible in (5.30). For simplicity, we suppress the $\sigma \cdot F$ OE of (5.30); but that dependence can be incorporated without difficulty, and the conclusions are unaltered. The effect of all the radiative corrections corresponding to non-perturbative fluctuations of the quantized electromagnetic field can now be seen by inserting (6.10) into the quenched version of (6.9):

$$\langle p|S'_c[A^{\text{ext}}]|p'\rangle = e^{\mathcal{D}_A}\langle p|G_c[A^{\text{ext}} + A]|p'\rangle|_{A\to 0}. \tag{6.11}$$

The essential operation is the linkage operator acting on the second line of (6.10), which yields

$$N' \int d[\Omega]\mathcal{F}[\Omega] \int d[\phi]e^{i\int_0^s ds' \phi_\mu f_\mu}$$

$$\cdot \exp\left[i\frac{g^2}{2}\int_0^s ds_1 \int_0^s ds_2 \phi_\mu(s_1)D_{c,\mu\nu}(\Xi)\phi_\nu(s_2)\right], \tag{6.12}$$

where

$$f_\mu(s') = \Omega_\mu(s') - gA_\mu^{\text{ext}}\left(\zeta(s') - 2\int_0^{s'}\Omega\right)$$

and

$$\Xi(s_1, s_2) = (s_1 - s_2)(p + p') - 2\int_{s_2}^{s_1} ds'' \Omega(s'').$$

The functional integral (FI) of (6.12) is now Gaussian, and produces

$$[N']^{\frac{1}{2}}\int d[\Omega]\mathcal{F}[\Omega]e^{-\frac{1}{2}\text{Tr}\ln K}$$

$$\cdot \exp\left[-\frac{i}{2}\int_0^s\int_0^s ds_1 ds_2 f_\mu(s_1)\langle s_1|(K^{-1})_{\mu\nu}|s_2\rangle f_\nu(s_2)\right],$$

or

$$[N']^{\frac{1}{2}}\int d[f]\mathcal{F}[f + gA^{\text{ext}}]e^{-\frac{1}{2}\text{Tr}\ln K} \cdot e^{-\frac{i}{2}\int f \cdot K^{-1} \cdot f} \cdot \exp\left\{-\text{tr}\ln\left[1 - g\frac{\delta A^{\text{ext}}}{\delta \Omega}\right]\right\},$$

$$\tag{6.13}$$

where $\langle s_1|K_{\mu\nu}|s_2\rangle = g^2 D_{c,\mu\nu}(\Xi)$. Equation (6.13) is a Gaussian-weighted FI over $\int d[\Omega]$, or over $\int d[f]$, which is sufficiently complicated that it cannot be evaluated explicitly. However, the map of (5.33) no longer appears, nor will the possibility of chaos which results from that map. Here, the sharp δ-functional has been replaced by a smoother, Gaussian-weighted integrand over a kernel defined by the radiative corrections; and however complicated the non-perturbative result may be, it is not the chaos of Section 6.1. Rather, it is a clear example of what has been termed[8] "environment-induced decoherence", as the radiative corrections remove the special coherence of the δ-functional, along with the map that can lead to chaos. It is straightforward to see that the same conclusions will be reached when any number of $G_c[A^{\text{ext}} + A]$ and $L[A^{\text{ext}} + A]$ factors are included, and hence the possibility of chaos stemming from vectorial interactions in QED (and, similarly, in QCD) is removed by the totality of radiative corrections.

An interesting explanation of this effect has been suggested in a recent paper by Dabaghian;[9] and while the method used is somewhat heuristic, and not mathematically rigorous, it is intuitive and well worth describing in some detail. Dabaghian first rephrases the Fradkin representation of $G_c[A]$ in terms of a functional integral over particle-like trajectories $x_\mu(s)$, with $v_\mu(s) = \mathrm{d}x_\mu(s)/\mathrm{d}s$, which are solutions to an effective, dynamical equation, $\frac{\mathrm{d}x_\mu}{\mathrm{d}s} = \frac{1}{m}P_\mu(s) - \frac{g}{m}A_\mu^{(0)}(x)$, that (depending on $A_\mu^{(0)}$) may or may not be chaotic. The quenched propagator, $S_c'[A^{(0)}]$, with full radiative corrections, is then exhibited as a FI over a new, vector field A_μ, which displays Gaussian-weighting [in the difference variable $g(A - A^{(0)})$] over a quantity that resembles the original $G_c[A^{(0)}]$, but is defined in terms of solutions to the dynamical system $\mathrm{d}x_\mu/\mathrm{d}s = \frac{1}{m}P_\mu(s) - \frac{g}{m}A_\mu(x)$. For the quantity $|S_c'[A^{(0)}]|^2$, Dabaghian then shows how the properties of these effective, dynamical systems can be compared and understood.

The behavior of each system can be analyzed in terms of trajectories of an equivalent phase-space point. There is, for $|S_c'|^2$, an FI to be estimated, whose major contributions arise from trajectories corresponding to phase-point motion on KAM tori of effectively-integrable systems. If the original $A^{(0)}$ system is chaotic, its phase point can display "Arnold diffusion",[10] as it wanders in-between the web of rational tori. According to this picture, a system will naturally use available tori in phase space to produce the most important contributions to a relevant FI; but if the number of conserved integrals-of-motion of that system is less than the number of degrees of freedom, a portion of its trajectories will wander over the phase space and display the "diffusion" associated with chaos.

But once the radiative corrections are included, the situation changes drastically. New, "virtual tori" appear, corresponding to motion defined in terms of A_μ, rather than $A_\mu^{(0)}$; and the important contributions to the new integral will be those corresponding to the effectively-integrable motion of the phase point on these new, virtual tori. Although the functions A_μ are "g-close" to the original, chaos-inducing $A_\mu^{(0)}$, in the sense of the Gaussian-weighted distribution, some of the systems characterized by A_μ are certainly integrable; even more, it is known that integrable trajectories are dense around each non-integrable trajectory, so that those A_μ which correspond to integrable systems are not rare. In the limit when the radiative corrections' coupling $g \to 0$, the Gaussian distribution simply yields the result corresponding to the original $A^{(0)}$ system, as all quantum field-fluctuations disappear, and so do the virtual tori. This means that the "white noise" of the radiative corrections in effect restructures the phase space, and in this way removes the original chaos of the $A^{(0)}$ system.

The interested reader is urged to refer to Dabaghian's paper for relevant, mathematical details. In the next section, we define a crude method for applying "radiative corrections" to a pair of simple, well-known, classical, chaotic systems; and observe that such "quantum fluctuations" do indeed tend to suppress the original chaos. And the way in which this happens, in the most striking example, is very much suggestive of a phase point initially displaying chaotic "wandering", and then suddenly finding a new, Dabaghian, multiply-periodic set of orbits, corresponding to subsequent, integrable motion.

6.3 Fluctuation-induced chaos suppression

Even though the chaos displayed by solutions to the differential equations (DEs) of this section have little to do with that generated by interactions with a vector-function $A^{(0)}$, the inclusion of external random and/or chaotic fluctuations – which mimic the above radiative corrections – can be performed in such a way as to correspond to the behavior of (macroscopic) quantum fluctuations. It should be emphasized at the outset that this is not an attempt to remove chaotic behavior in a sophisticated manner, such as that suggested by Auerbach *et al.*,[11] wherein one tries to anticipate and direct the motion of a chaotic phase-space point; rather, the method suggested here goes best by the name "brute force", and leaves various questions unanswered. (For example: what is the form of the new, integrable Hamiltonian?) Nevertheless, one observes behavior which is strongly suggestive of Dabaghian's analysis; and which, therefore, deserves a rigorous, mathematical proof, or disproof.

Imagine a DE whose solution $x(t)$ displays obvious chaotic behavior. We insert into the DE, in an appropriate fashion, "oscillatory" time dependence, say $\cos(t\omega)$, where ω is first chosen as a random input function, $\omega(t)$. Then, in the numerical integration of the resulting equations, there will be uncontrollable fluctuations $\Delta\omega$ at each step, which roughly corresponds to the insertion of quantum fluctuations of a system satisfying an $\hbar \sim 1$ version of the Heisenberg Uncertainty Principle. Essentially, large fluctuations are associated with small time-intervals, while the large-time behavior of the system should correspond to averaged ω-variations of smaller magnitude than the individual $\Delta\omega$ fluctuations.

The result of this insertion is a clear diminution of the region of phase space in which chaotic motion takes place. If one goes a step further, and replaces the external $\omega(t)$-fluctuations in the original DE for $x(t)$ by internal x-fluctuations, one finds that, after an initial period of "diffusion", the chaos is apparently removed, as the phase-point's motion appears to become integrable. This suggests that the phase point "wanders" just until it finds a nearby, Dabaghian virtual

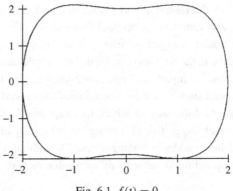

Fig. 6.1 $f(t) = 0$

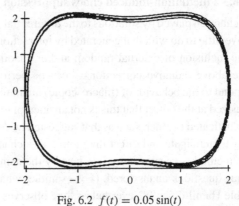

Fig. 6.2 $f(t) = 0.05 \sin(t)$

torus, about which it then performs integrable motion. Similar effects occur for a second, well-known system, whose original chaos is of a completely different origin.

In detail, consider the forced Duffing model, described by the DE $\ddot{x} + (x^2 - 1)x = f(t)$. For $f(t) = 0$, and the initial conditions (which are not relevant to the procedure) $x(0) = 2$, $\dot{x}(0) = 0$, one has a well-known problem with solution given by the Jacobi elliptic function corresponding to positive energy, as shown in Fig. 6.1. Chaotic motion is most simply achieved by choosing $f(t)$ oscillatory, beginning with a small magnitude, as in Fig. 6.2, and proceeding to an amplitude of reasonable magnitude, e.g., $f(t) = 5 \sin(t)$, as in Fig. 6.3.

The phase portrait of Fig. 6.3 leaves no doubt that this is now a chaotic system. Imagine another system with a random coordinate, or perhaps a system with a phase portrait similar to that of Fig. 6.3, and denote its coordinate by $y(t)$. Then, the coupling of this external, chaotic "y-system" can be accomplished by inserting a certain measure of y-dependence in $f(t)$, replacing the latter by

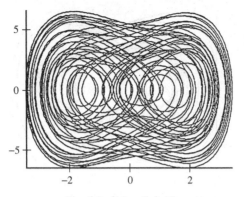

Fig. 6.3 $f(t) = 5\sin(t)$

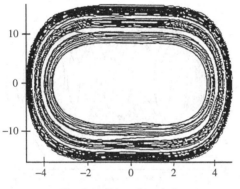

Fig. 6.4 $f(t) = 5\cos(ty)$

$f(t \cdot H(y))$, where $H(y)$ may be chosen in a variety of ways, of which perhaps the simplest is illustrated in Fig. 6.4, $H(y) = y$. Because $H(y)$ is chaotic, and is therefore in some measure random, one can think of $H(y)$ as a frequency $\omega = E/\hbar$, whose random variations correspond to quantum-like energy fluctuations of "classical" size. One sees that the phase point is essentially confined to the darker, band-like regions of phase space, within which the motion appears chaotic.

Since the motion within the bands is still chaotic, one may ask if it is necessary to use an external chaotic signal, $H(y)$; why not simply use the chaotic self-interaction $H(x)$? Accordingly, we again begin with the chaotic choice $f(t) = 5\sin(t)$ of Fig. 6.3, and then replace this by $f(tH(x))$, with $H(x) = 1 + cx$, with increasing values of c as shown in Figs. 6.5, 6.6, and 6.7. Now what happens is clear, and quite attractive. After a few chaotic-seeming excursions, the previous band-motion is replaced by a combination of small, rapid oscillations superimposed on a (relatively) slowly varying background.

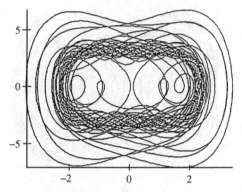

Fig. 6.5 $f(t) = 5\sin(t[1 + 0.03x])$

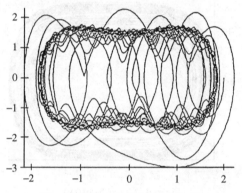

Fig. 6.6 $f(t) = 5\sin(t[1 + 0.3x])$

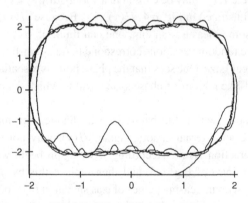

Fig. 6.7 $f(t) = 5\sin(t[1 + 3x])$

As the parameter c is increased, the rapid oscillations decrease in size, while the background figure resembles more and more closely the original Jacobi solution without forcing. It is the combination of initial, wild excursions, plus an abrupt settling down to motion which certainly looks integrable (rapid oscillations on a slowly varying background), that brings Dabaghian's "virtual tori" to mind, and gives one the hope that his intuitive analysis can be made mathematically precise.

A similar, self-interaction effect can be seen for the chaotic system defined by the pair of coupled-oscillator equations

$$\ddot{x} + y^2 x = 0, \qquad \ddot{y} + x^2 y = 0,$$

which can be re-written in first-order form as

$$\dot{x} = u, \qquad \dot{u} = -y^2 x, \qquad \dot{y} = v, \qquad \dot{v} = -x^2 y. \tag{6.14}$$

These equations follow from the variation of the Lagrangian

$$L = \frac{1}{2}(\dot{x}^2 + \dot{y}^2) - \frac{1}{2}x^2 y^2,$$

which generates a conserved Hamiltonian as the only constant of the motion. Instantaneous Lyapunov exponents are easily calculated, and one can show that there always exists one which is positive, indicative of instability, and in this case, chaos. When $(\dot{x}, x)$ and $(\dot{y}, y)$ phase portraits are calculated, that chaos takes the form of unpredictable oscillations of both variable pairs such that, e.g., the x-motion resembles a horizontal oval, while the y-motion resembles a vertical oval; and, then, unpredictably, the two motions are suddenly interchanged.

Now, let us add a "quantum mechanical" coupling to these equations, which can be done in a variety of ways; e.g., by adding to the RHS of the second and fourth equations of (6.14) the terms $f_0 \cos(tx)$ and $f_0 \cos(ty)$, respectively. Immediately there is a change in the motion, with, e.g., the x-motion continuing to be mainly vertical, while the y-motion takes the form of very slow horizontal motion; it is as if chaos is here being removed, or at least suppressed, because the y-motion is almost fixed in place, while the x-motion appears integrable.

To a physicist, from these two examples it is intuitively clear that such "quantum-mechanical" coupling can act to remove chaos; to a mathematician, the problem has just barely been formulated. But the questions of if and how this procedure does work are surely most interesting. The above results are the antithesis of previous examples where periodic forcing is used to remove the chaos originally present in a Lorenz model.[12] They do have a certain similarity to the biological observations of Halloy et al.,[13] where the coupling of chaotic cells to periodic cells tends to suppress the original chaos (as the combined systems bifurcate towards a different regime of total chaos). Here, however, we are basically considering the interaction of two chaotic systems, or of the self-interaction of a single chaotic system; and we find that the chaos of the affected system becomes limited in the sense that the chaotic regions of phase-space tend to be squeezed into bands, or, for the self-interacting system, that the chaotic system is apparently changed by the quantum-mechanical fluctuations into one strongly resembling an integrable system[14].

Notes

1 See, for example, *Engineering Applications of Dynamics of Chaos*, edited by W. Szemplinska-Stupnicka and H. Troget, Springer-Verlag, Wien, 1991.
2 See, for example, L. E. Reichl, *The Transition to Chaos*, Springer-Verlag, Berlin, 1992, and the very many references cited therein.
3 See Note 5, Chapter 4.
4 An interesting variant to the particle-field decomposition presented here is provided by recent work of the Duke group, who solve the classical Yang–Mills field equations on a lattice, and find chaotic behavior illustrated by an array of positive Lyapunov exponents: B. Müller and A. Trayonov, *Phys. Rev. Letters* **23** (1992) 3387; C. Gong, *Phys. Lett. B* **298** (1993) 257 and *Phys. Rev. D* **49** (1994) 2624; C. Gong, S. G. Matinyan, B. Müller, and A. Trayonov, *Phys. Rev. D* **49** (1994) R607.
5 H. M. Fried, in *Proceedings of the Vth Rencontre de Blois "Chaos and Complexity"*, Blois, France, edited by J. Tran Thanh Van, Éditions Frontières, Gif-sur-Yvette, 1993.
6 B. Delos, S. K. Knudsen, and D. W. Noid, *Phys. Rev. A* **30** (1984), and references quoted therein.
7 The author is indebted to Walter Craig for this discussion and demonstration.
8 H.-T. Elze, *Nucl. Phys. B* **436** (1995) 213; W. Zurek and J. P. Paz, Los Alamos report No. LA-UR-94-927, and other references quoted therein.
9 Y. Dabaghian, *Phys. Rev. Letters* **77** (1996) 2666. A detailed discussion of many of these topics may be found in the Ph.D. thesis of this author, "Non-Perturbative Aspects of Quantum Chaos Theory", Physics Department, University of Rhode Island, 1999.
10 V. I. Arnold, *Russ. Math. Survey* **18** (1963) 1885.
11 D. Auerbach, C. Grebogi, E. Ott, and J. A. Yorke, *Phys. Rev. Letters* **69** (1992) 3479.
12 G. M. Guidi, J. Halloy, and A. Goldbeter, "Chaos Suppression by Periodic Forcing", in *Proceedings of the Vth Rencontre de Blois*, see Note 5.
13 J. Halloy, H. Y. Li, J. L. Martiel, B. Wurster, and A. Goldbeter, *Phys. Letters A* **151** (1990) 33; and Errata, *Phys. Letters A* **159** (1991) 442.
14 The author thanks Prof. G. Guralnik for many discussions and his help with matters computational; Prof. S. Matinyan for pointing out the existence of chaos in the model of (6.14); and Dr. W. Becker for his kind aid in surveying the literature. Numerical integrations described above were carried out using the fifth-order, Runge–Kutta ODE algorithm, with adaptive-stepwise control, as given by H. J. Korsch and H.-J. Jodl, in the book *Chaos*, Springer-Verlag, Berlin (1994). These results were first presented at the June 1996 *Workshop on QCD: Collisions, Confinement, and Chaos*, at the American University of Paris, with proceedings published by World Scientific in 1997.

7

Infrared approximations

No book on Green's functions can omit a discussion, however brief, on the subject of Infrared (IR)/no-recoil/Block–Nordsieck models, which have played a central role in the description of those processes for which there is a clear, physical separation between large and small frequencies. The most famous example is probably the removal of all IR divergences in QED, and the way in which this may most simply be illustrated clearly exhibits the power of functional methods.[1] A second example, which has hardly received the attention it merits, is the propensity of virtual soft-photons to damp processes involving large momentum transfers,[2] a subject which will be briefly described in Section 7.2.

In modern language, the true role of soft photons, or very low-frequency photon fluctuations, was anticipated by Bloch and Nordsieck in their seminal paper[3] of 1937; but not until almost two decades had passed was a proof given of the cancellation of all IR divergences for any scattering process in QED. The latter constructions, by Yennie, Frautschi, and Suura,[4,5] were performed by identifying and extracting the IR divergent terms in every order of perturbation theory, and then summing over them all to obtain a final $|\text{amplitude}|^2$ explicitly free of all IR divergences. Shortly afterwards, Schwinger[6] and Mahanthappa[7] invented a functional method for the direct calculation of probabilities, in which the IR difficulties never appear. Following in these footsteps, the author has given explicit functional constructions[1] in which the IR divergences of QED are shown to cancel, leaving behind the residual and experimentally-important "radiative corrections";[4] and the starting point for those functional calculations was the construction of the electron propagator $G_{\text{BN}}[A]$ in the limit where the frequency k of $\tilde{A}_\mu(k)$ is much smaller than the relevant electron momenta p: $|k^2| \ll |p \cdot k|$. Such approximations were originally defined without the aid of the Fradkin representation, and for completeness and subsequent usage in an eikonal context, Section 7.1 reproduces the corresponding Svidinski/Symanzik construction;[8] for the application of this result to the functional elimination of

107

IR divergences, the reader is referred to the books of Note 1, as well as for a brief description of how and when IR divergences are removed in QCD.

Section 7.2 deals with the damping of large-momentum-transfer processes by soft, virtual photons; and a use for this property is displayed by recalling simple, two-parameter fits to well-known electromagnetic-vertex form factors, and wide-angle scattering amplitudes.[9] Even though this material is some three decades old, it is, in the author's judgement, worth mentioning because of the very real possibility that analogous forms (with one important difference) in QCD may be able to shed some non-perturbative light on spin-polarization experiments at high energies and large momentum transfers; this will be briefly discussed in Sections 8.1 and 8.2. In Section 7.3, the forms of $G_{BN}[A]$ are used to set up eikonal representations of scattering amplitudes in particle physics, as an introduction to Section 8.4; and, finally, in Section 7.4, IR approximations and their rescaling corrections for non-scattering problems – another subject deserving of more attention than it has received – are briefly discussed.

7.1 The Block–Nordsieck approximation

We work directly with the fermion $G_c[A]$ of QED, satisfying

$$[m + \gamma \cdot (\partial_x - ig A(x))]G_c(x, y|A) = \delta^{(4)}(x - y), \tag{7.1}$$

or, using our previous, formal notation,

$$[m + \gamma \cdot (\partial - ig A)]G_c[A] = 1. \tag{7.2}$$

This $G_c[A]$ is to describe a charged fermion emitting or absorbing real or virtual photons (or any NVM quanta to which the fermion field is coupled); and since 4-momentum conservation must hold in every such emission or absorption, the fermion motion must reflect this conservation. However, if the fermion momenta are much larger than the boson momenta emitted or absorbed, the fermion hardly suffers any recoil in each such event. This is just the IR limit of QED, where only very small photon 4-momenta are considered; it is also a useful limit that can be applied in other situations, when the fermion 4-momenta are large, and the boson mass need not be zero.

In all such cases, the essential physical approximation is the lack of fermion recoil. The mathematical expression of this limit is the replacement of the Dirac γ_μ matrices by c-number constants, $-iv_\mu$, representing an "averaged" fermion 4-momentum, unchanged by multiple soft emission and/or absorption. Here, v_μ is to represent the 4-velocity, $v_\mu = p_\mu/m$, of that (incoming or outgoing) fermion which appears, e.g., in an appropriate place in the functional construction of a scattering amplitude; in each case, $v^2 = -1$, and in that fermion's rest frame, $v_0 = +1$. The basic DE satisfied by $G_c[A]$ is now changed to read

$$[m - iv \cdot (\partial - ig A)]G_{BN}^{(v)}[A] = 1, \tag{7.3}$$

and may be solved exactly for any $A(x)$. Note that (7.3) is a first-order DE – basically non-relativistic, in spite of the 4-dimensional notation, because the antiparticle pole in the k_0 plane is missing – and its solutions may be expected to be either "retarded" or "advanced"; our construction yields the former.

The solution for $G_{BN}^{(v)}[A]$ proceeds in a manner similar to (but much simpler than) the Fradkin construction of Chapter 3; one writes the inverse of (7.3) in the form

$$G_{BN}^{(v)}[A] = i \int_0^\infty ds \, e^{-is[m - iv \cdot (\partial - ig A)]} = i \int_0^\infty ds \, e^{-ism} \cdot e^{-sv \cdot \partial} F(s), \tag{7.4}$$

where $F(s) = e^{sv \cdot \partial} \cdot e^{-sv \cdot (\partial - ig A)}$.

To determine the explicit form of $F(s)$ one calculates the variation of its $\langle x|$ projection with respect to s,

$$\langle x| \frac{\partial F}{\partial s} = \frac{\partial}{\partial s} \langle x|F(s) = igv \cdot A(x + sv)\langle x|F(s),$$

so that

$$\langle x|F(s) = e^{ig \int_0^s ds' v \cdot A(x + s'v)} \langle x|,$$

and

$$G_{BN}^{(v)}(x, y|A) = i \int_0^\infty ds \, e^{-ism} \delta^{(4)}(x - y - sv) e^{ig \int_0^s ds' v \cdot A(x - s'v)}. \tag{7.5}$$

Note that $G_{BN}[A]$ is a retarded function in any Lorentz frame, and hence the $L[A]$ constructed from (7.5) must vanish.

In order to proceed from n-point functions to S-matrix elements, certain operations upon the former are necessary; in particular, the requirement of mass-shell amputation[8] is paramount. Here, one must, e.g., take $G_c(u, x|A)$, where x corresponds to an incoming configuration-space variable, and operate upon this G_c with the Dirac operator $[m - \gamma \cdot \partial_x]$ or $[m + i\gamma \cdot p]$, then calculate its Fourier transform $\int d^4x \exp[+ip \cdot x]$, and then go to the mass shell (which, between Dirac spinors, means the algebraic statement equivalent to the limit: $m + i\gamma \cdot p \Rightarrow 0$). For our BN model, this is simply the combination

$$\int d^4x e^{ip \cdot x} G_{BN}^{(v)}(u, x|A)(m + v \cdot p)|_{m.sh.},$$

which, after an integration by parts, yields

$$e^{ip \cdot u} \cdot \exp\left[ig \int_0^\infty ds\, v \cdot A(u - sv)\right]. \tag{7.6}$$

Similarly, if y denotes an outgoing particle, of momentum p', one obtains

$$\int d^4y\, e^{-ip' \cdot y}(m + v' \cdot p') G_{BN}^{(v')}(y, w|A)|_{m.sh.}$$

$$= e^{-ip' \cdot w} \cdot \exp\left[ig \int_0^\infty ds\, v' \cdot A(w + sv')\right]. \tag{7.7}$$

Both (7.6) and (7.7) are quite useful relations, for both close-to-forward and wide-angle scattering, as will be seen in the following sections.

7.2 IR damping at large momentum transfers

Imagine that a charged particle is scattering in a weak, external field, for which the Green's function $G_c(x, y|A + A^{ext})$ is needed, with $A(z)$ denoting the vector potential of the fluctuating ("soft") electromagnetic field. To first order in A^{ext}, one writes

$$G_c(x, y|A + A^{ext}) \simeq G_c(x, y|A) + ig \int d^4z\, G_c(x, z|A)\gamma \cdot A^{ext}(z) G_c(z, y|A), \tag{7.8}$$

where the first RHS term of (7.8) may be dropped, since the reaction we wish to study requires at least a first-order dependence in A^{ext}, in order to achieve the desired wide-angle scattering. We now specialize to the case of soft photons, replacing the remaining RHS term of (7.8) by

$$ig \int d^4z\, G_{BN}^{(v')}(x, z|A)\gamma \cdot A^{ext}(z) G_{BN}^{(v)}(z, y|A), \tag{7.9}$$

where $v = p/m$ and $v' = p'/m$; this corresponds to a (semi-classical) electron scattering from y to x, with a corresponding momentum change from p to p'. From Section 2.5, one sees that the quantum fluctuations of the electromagnetic field for this process – and here we adopt the "quenched' simplification, neglecting closed fermion loops, with $L[A] \to 0$, $\langle 0|S|0 \rangle \to 1$ – are expressed by the linkage operator $\exp[-(i/2) \int \frac{\delta}{\delta A} D_c \frac{\delta}{\delta A}]$, where $D_c(k) = [k^2 + \mu^2 - i\epsilon]^{-1}$ is the free, causal photon propagator, in (for simplicity) the Feynman gauge,

$D_{c,\mu\nu} = \delta_{\mu\nu}D_c$, and μ is a small photon mass inserted to regulate any untoward IR divergences.

All the A-dependence here lies in the exponential factors of (7.6) and (7.7); and the linkage operation (after which $A \to 0$) is immediate, generating

$$\exp\left[\frac{i}{2}g^2v^2 \int_0^\infty ds_1 \int_0^\infty ds_2 D_c([s_1 - s_2]v)\right]$$
$$\cdot \exp\left[\frac{i}{2}g^2v'^2 \int_0^\infty ds_1 \int_0^\infty ds_2 D_c([s_1 - s_2]v')\right]$$
$$\cdot \exp\left[ig^2(v \cdot v') \int_0^\infty ds_1 \int_0^\infty ds_2 D_c(s_1v + s_2v')\right]. \qquad (7.10)$$

The first two RHS terms of (7.10) are associated with the self-energy structure of the incoming and outgoing electrons, and a similar factor appears in the proper definition of the S-matrix which relates the Green's function to the desired probability amplitude; the combination, treated in detail in HMF#1 and HMF#2, produces the multiplicative factor

$$\exp\left[\frac{ig^2}{2(2\pi)^4}\right]\frac{d^4k}{k^2 + \mu^2 - i\epsilon}\left(\frac{p}{k \cdot p + i\epsilon} - \frac{p'}{k \cdot p' + i\epsilon}\right)^2, \qquad (7.11)$$

where both p and p' are assumed to be on their mass shells, $p^2 = p'^2 = -m^2$. For small, virtual, photon momenta, (7.11) multiplied by γ_μ represents the soft-photon renormalized vertex function to all orders in g^2.

In the limit of $\mu \to 0$ the exponential of (7.11) displays an IR divergence, which, in QED, will damp to zero every amplitude constructed from it, unless the possibility of scattering and simultaneously producing an infinite number of real, soft photons is calculated; and in that case all IR divergent factors in the expression for the probability of this process exactly cancel. If one studies a massive NVM theory, so that one need not consider soft-photon production in order to remove the IR divergences, then the evaluation of this vertex function is of considerable interest, for it provides a significant damping at large momentum transfer, real or virtual. As it stands, however, (7.11) requires a UV cut-off, simply because all virtual photon momenta considered in its construction were initially assumed soft. Other forms, used in the classic papers of Notes 4 and 5, have replaced the combination

$$\left(\frac{p}{k \cdot p} - \frac{p'}{k \cdot p'}\right)^2$$

of (7.11) by

$$\left(\frac{2p+k}{k^2 + 2p \cdot k} - \frac{2p'+k}{k^2 + 2p' \cdot k} \right)^2,$$

which resembles the ordinary perturbative functions – and does not require a UV cut-off in (7.11) – and which will lead to an extra factor of $\ln(q^2/m^2)$ in the exponential of (7.11) for $q/m \gg 1$, e.g.,

$$\Gamma_\mu \sim \gamma_\mu \exp[-\lambda \ln^2(q^2/\mu^2)], \quad \lambda > 0,$$

a result originally obtained by Jackiw[10] as the sum of the leading $\ln(q^2)$ dependence in every order of perturbation theory for the vertex function in massive-photon QED.

The required UV cut-off of (7.11) can be conveniently arranged by inserting under the k-integral the factor $\exp[-i\alpha k^2]$, where the variable α is treated as a real, positive number, and at the end of the calculation is continued to an imaginary value according as $\alpha \to -i\mu_c^{-2}$, where μ_c represents a real, positive cut-off with dimensions of mass. With this prescription, the exponential of (7.11) may be easily evaluated to yield $\gamma F(t)$, where

$$F(t) = t \int_{4m^2}^{\infty} \frac{dt'}{t'} \frac{1}{(t'-t)} \left(1 - \frac{2m^2}{t'} \right) \left(1 - \frac{4m^2}{t'} \right)^{-1/2}, \qquad (7.12)$$

and

$$\gamma = \frac{g^2}{8\pi^2} \int_0^{\infty} \frac{db}{b + \mu_c^{-2}} e^{-b\mu^2} \simeq \frac{g^2}{8\pi^2} \ln\left(1 + \mu_c^2/\mu^2 \right).$$

One pleasant feature of this method of approximation is that the three constants needed to specify the soft exchanges, g, μ, and μ_c, coalesce into the single constant γ, which may be treated as a parameter to be determined by experiment. For negative t, $F(t)$ is real and negative,

$$F(t) = 1 - \frac{2x+1}{[x(x+1)]^{1/2}} \cdot \ln\left(\sqrt{x} + \sqrt{x+1} \right), \qquad x = -\frac{t}{4m^2} > 0,$$

with limiting forms $t/3m^2$ and $-\ln(-t/m^2)$ for small and large $-t$, respectively. If m denotes the nucleon mass, a very good approximation for all $-t$ (and $|t|$ in GEV^2) is given by $F(t) = -\ln(1 + 0.4|t|)$; and this has been used to give a two-parameter fit to the nucleon electromagnetic form factors[9] which is considerably better than the more common, empirical, two-parameter dipole-fit formula.

For positive argument, one can show that Re $F(s) = F(4m^2 - s)$, a relation which becomes part of quality, wide-angle fits to nucleon–nucleon scattering at large momentum transfers. Here, one has an exponential factor of the form (7.6) or (7.7) for each fermion leg entering or leaving the reaction, and the soft linkages across each pair of legs produce damping of the scattering amplitudes, leading in the appropriate limits to experimentally-observed behavior: for $s \rightarrow \infty$ and fixed (but large) $|t|$, to the Chou–Yang–Wu observation;[11] and for large s, $-t$, $-u$, to qualitatively correct dependence on the Krisch variable, ut/s.[12] In brief, at least for energies below which one must seriously take the quark/gluon content of matter into account, these IR summations provide a simple way of understanding a wide range of experimental data.

There is another, and somewhat tantalizing, theoretical idea which may be worth mentioning. Ordinary perturbation theory, e.g., of massive NVMs coupled to fermions, is plagued with UV logarithmic divergences which happily disappear when renormalization – the passage, as Schwinger used to say, from the field to the particle point of view – is taken into account. Some years ago, such divergences were taken seriously, with Källén[13] arguing that at least one of the renormalization constants of QED was infinite, while Baker and Johnson[14] tried to develop a scheme that would tie the vanishing of these UV divergences to the experimental value of the renormalized, fine-structure constant. Now, of course, we know that QED cannot be considered independently of the weak interactions, while many believe that the weak, strong, and electromagnetic interactions are just broken-symmetry forms of a grand unified theory. Whatever one believes, we still characterize our theories in terms of perturbative expansions: is it renormalizable or not? But even for the simplest of nontrivial, 4-dimensional theories, order-by-order the renormalization constants diverge.

Our experience above of summing over soft, NVM radiative corrections suggests that a modified perturbation expansion (MPE) might be possible: not an expansion in powers of an unrenormalized coupling, but rather an expansion in "powers" of boson propagators which contain only large momenta, with the soft linkages counted as the zeroth part of the expansion. Could such a MPE generate a sequence of terms each of which is finite before renormalization?

A simple example will make this clear. Consider the simplest, fermion, self-energy graph; but for one vertex – as suggested by the corresponding Dyson–Schwinger equation – insert the soft-photon, renormalized, vertex function under the integrand. It is not difficult to see that the IR damping obtained above for large k^2 when both fermion legs were on their mass shell is still present – although changed somewhat in form – if only one fermion leg is on its mass shell; and since the integrand without the soft-damping is only logarithmically divergent, any extra convergence factor at large k will be sufficient to render the entire integral finite. This suggests a scheme in which all soft radiative corrections are introduced and retained as the very first step of an MPE, while higher-order corrections are obtained by expanding in the number of "hard" virtual quanta exchanged.

Any such scheme must preserve the gauge-independence of $L[A]$, or $L[F]$. Any function of $F_{\mu\nu}$ is, by definition, gauge invariant in QED; but in QCD, where $L[F]$ is rigorously gauge invariant, any approximation to L must also display that property. What is required here is invariance under local, configuration-space transformations; but the IR approximation begins with a local restriction in momentum space. How can any IR prescription, which is then non-local in configuration space, be compatible with invariance under local, configuration-space transformations? In fact, this is possible; and three distinct methods for preserving the QCD gauge invariance of the final results are described in HMF#2.

The same source describes an IR approximation to $L[F]$ in QED, which is a generalization of Schwinger's solution for the special case when the $F_{\mu\nu}$ are constant; there the frequency components of $F_{\mu\nu}$ are not zero, but are carefully arranged to be (qualitatively) less than the loop momenta. For technical reasons, a second, "multipole" approximation is here required and introduced. From this L one constructs, by functional differentiation, the photon propagator containing all possible "soft" radiative corrections, a quantity which superficially appears to be finite, and which would enter into the subsequent "hard" corrections. The word "superficially", however, is most appropriate, because of the need to blend gauge invariance with such soft-photon corrections in the lowest and in every subsequent order. It is not clear, as least to the present author, that this can actually be done; but the damping of hard momenta by sums over virtual soft momenta suggests that this approach should be tried, for it could well lead to finite QFT, even for interactions which are presently believed too "non-renormalizable" to be attempted by ordinary perturbation theory.

7.3 Eikonal scattering amplitudes in particle physics

A brief but self-contained, functional derivation of eikonal scattering amplitudes may be found in HMF#1 and HMF#2, and need not be repeated here; this subject requires a certain familiarity with the S-matrix, and the way in which its elements may be expressed in terms of appropriate, mass-shell-amputated, n-point functions of QFT. Perhaps the most comprehensive collection of eikonal references may be found in the book by Cheng and Wu,[15] whose later chapters describe the perturbative calculations which have been made for the eikonal function in QCD. The intent of the present remarks is more in the nature of a survey of what has been done, rather than a derivation of those results; and to set the stage for an explicit and partially complete eikonal for high-energy scattering in the so-called "multiperipheral model", as derived in the following chapter and involving a soluble and most relevant Green's function.[16]

Aside from multiplicative and renormalization constants, the essence of the connection between Green's functions and corresponding S-matrix elements, in the limit of large scattering energy and small momentum transfer, lies in the exponential factors of (7.6) and (7.7). When the scattering fermion couples to an NVM, and the exchange of arbitrary numbers of NVMs between a pair of scattering fermions is desired, eikonalization of the amplitude takes place[17] such that the scattering amplitude is expressed in terms of an eikonal function

$$T(s,t) = \frac{is}{2m^2} \int d^2b\, e^{i\mathbf{q}_\perp \cdot \mathbf{b}}\left[1 - e^{i\chi(s,b)}\right], \tag{7.13}$$

where $-t = q^2$ and $s =$ (total CM energy)^2. The eikonal is a function of impact parameter b, and of s; and for the simple case of the exchange of an arbitrary number of virtual NVMs between the scattering fermions, one finds the result

$$i\chi_1(b,s) = -\frac{ig^2}{2\pi}\gamma(s)K_0(Mb), \qquad \gamma(s) = \frac{(s-2m^2)}{\sqrt{s(s-4m^2)}}, \tag{7.14}$$

where g is the fermion-NVM coupling constant, and M denotes the NVM mass; to this eikonal there correspond the graphs of Fig. 7.1. At high energies, the invariant differential cross section is given by $d\sigma/dt = (m^4/\pi s^2)|T|^2$, while the total cross section is given by

$$\sigma_{\text{TOT}}(s) = 2\text{Re} \int d^2b\left[1 - e^{i\chi(s,b)}\right]. \tag{7.15}$$

Equations (7.13) and (7.15) are generic results, for any eikonal function, while (7.14) denotes the eikonal built from virtual NVM exchange between the scattering fermions.

One should understand, at least qualitatively, the straightforward, non-perturbative steps involved in the passage from (7.6) and (7.7) to (7.13) and (7.14). These steps consist of the following operations:

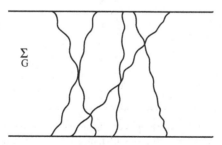

Fig. 7.1 The sum of all virtual NVMs exchanged between a pair of scattering fermions.

(i) Perform mass-shell amputation (msa) not on each Green's function that comprises T, but rather on each Green's function obtained by first calculating either $\partial T/\partial g^2$ or $\partial^2 T/\partial g_1 \partial g_2$, where $g_{1,2}$ are the couplings of the respective fermions to the NVMs. The reason for this indirect operation is that each Green's function then needs but one msa operation on one of its coordinates, and passage may at that time be made to the BN forms of (7.6) and (7.7). At the very end of the calculation, one must integrate over the couplings to achieve $T(g)$.

(ii) Retain, as the only explicit q-dependence, the difference of p and p' in the $\exp[ip \cdot z]$ factors of each fermion line, so that each fermion's A-dependence takes the form

$$e^{i\int d^4 w f_{I,II}^\mu(w) A_\mu(w)}, \quad f_{I,II}^\mu(w) = g p_{I,II}^\mu \int_{-\infty}^{+\infty} ds\, \delta(w - x_{I,II} + s p_{I,II}). \quad (7.16)$$

(iii) Perform the linkage operations corresponding to the exchange of all possible NVMs between the two fermions. Because the A-dependence of (7.16) is the exponential of a linear form, the linkage operation is immediate, and yields the eikonal of (7.14).

It should be noted that total cross section calculated for this eikonal from (7.15) becomes a constant, independent of s, as $s/m^2 \to \infty$, a fact which can be understood physically by the observation that the only inelastic graphs of this model are those of bremsstrahlung, and the latter always vanish for zero momentum transfer.

The important observation made by Cheng and Wu[18] was that, in massive NVM QED, there is another type of inelastic process which can contribute to inelastic production at small $|t|/s$, the so-called "multiperipheral" graphs pictured in Fig. 7.2, which, by unitarity, correspond to the "inelastic shadow" graphs of Fig. 7.3, graphs that must diminish the elastic amplitude if such

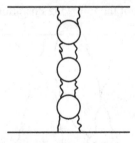

Fig. 7.2 An example of the tower graphs of Cheng and Wu. It should be understood that sums are to be performed over all possible numbers of closed fermion loops, and that all possible numbers of such towers are to be exchanged.

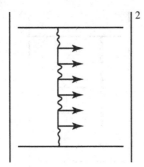

Fig. 7.3 A representation of the absolute square of the production amplitude obtained by calculating the absorptive part of the graph of Fig. 7.2.

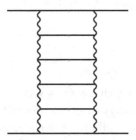

Fig. 7.4 Scalar mesons exchanged between virtual NVMs, as a variant of the Cheng–Wu model.

inelastic production increases as energies increase. In fact, there is a phase-space factor proportional to $\ln(s/m^2)$ for the probability of each fermion pair to be produced in this way, which suggests that the graphs of Fig. 7.3 are the relevant graphs to consider at high energies. Such "tower graphs" were calculated by Cheng and Wu, and yielded an eikonal of the form

$$i\chi_2(s, b) = -as^\alpha e^{-\mu b}, \tag{7.17}$$

with a, α, μ constants, which leads directly to the estimate that $\sigma_{\text{TOT}}(s) \sim \ln^2 s + \cdots$ in the limit of very high energies.[18]

Shortly after this observation, it was pointed out by various authors[19] that another, and simpler form of multiperipheral interaction generates a very similar result; this appears if scalar particles are exchanged between the NVM pairs, which are themselves exchanged between the scattering fermions, as in Fig. 7.4. Again, the unitarity "shadow" corresponds to inelastic graphs of the form shown in Fig. 7.5, where the phase-space of each scalar particle emitted contributes a factor of $\ln(s/m^2)$. The corresponding eikonal, constructed from the graphs of

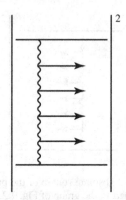

Fig. 7.5 A representation of the absolute square of the production amplitude obtained by calculating the absorptive part of the graph of Fig. 7.4.

Fig. 7.5, takes the form

$$i\chi_2 \simeq -\frac{a}{\ln(s/s_0)}(s/s_0)^{\alpha}e^{-\mu^2 b^2/\ln(s/s_0)}, \tag{7.18}$$

with a, α, μ, s_0 constants, which leads to a very similar elastic amplitude, and to the same form of high-energy $\sigma_{\text{TOT}}(s)$ as that of Cheng and Wu.

A functional description of all the eikonal graphs, including those not considered by Cheng and Wu and successors, was formulated by the author,[1] and can be expressed in simple functional language. One begins by considering fermions coupled to NVM fields, and the latter coupled to "scalar pion" (π) fields; and one then organizes the functional pieces that are involved in the desired scattering amplitude, discarding all terms which correspond to more structure than that of NVMs exchanged between a pair of scattering fermions, with all possible virtual π exchanges between all possible NVMs. In analogy with the forms quoted above, the NVM interactions eikonalize, carrying with them the composite substructures corresponding to multiple π exchange between all possible NVMs, with a resulting expression for the eikonal

$$e^{i\chi} = \exp\left[-\frac{i}{2}\int\frac{\delta}{\delta\pi}D_{\text{c}}\frac{\delta}{\delta\pi}\right]\cdot\exp\left[i\int f_{\text{I}}\cdot\bar{\Delta}_{\text{c}}[\pi]\cdot f_{\text{II}}\right]\Bigg|_{\pi\to 0}, \tag{7.19}$$

where $\bar{\Delta}_{\text{c}}(x,y|\pi)$ denotes the propagator of a NVM of mass M in the presence of a (fictitious) field $\pi(z)$, D_{c} is the free propagator of a scalar pion of mass μ, and the $f_{\text{I,II}}^{\mu}$ denote the classical currents of the two fermions. The only modification of (7.19), simple to perform in this functional context, will be to discard all terms which correspond to self-energy structure of virtual π emission and absorption along each NVM; the only graphs one wishes to retain are to correspond to virtual π exchange between NVMs, and in all possible ways. We shall return to (7.19) in the next chapter, with the aid of a new expression for a simplified $\bar{\Delta}_c[\pi]$ which contains the essential Physics of this model.

7.4 IR approximations and rescaling corrections to non-linear ODEs

IR approximations to non-scattering problems such as the $L[A]$ of Potential Theory and QFT, to the motion of fluids governed by the Navier–Stokes and/or Euler DEs, and to integrable, second-order ODEs have been presented, with references, in the later chapters of HMF#2. The IR method introduced was, firstly, used to discuss a suitably-modified Green's function for each problem, and then to perform an appropriate IR approximation designed to capture all (or almost all) of the solutions' relevant, long-distance, low-frequency behavior. Then, for non-linear ODEs, global "glissando" rescaling corrections (GRC) were devised to insert into those IR approximations some of the higher-frequency dependence which had initially been discarded. The purpose of this section is to simplify that IR/ODE analysis,[20] illustrated most simply by the Duffing ODE, and in particular to show explicitly how the first GRCs to the IR model correspond to a significant improvement in accuracy obtained in a simple way. Thus, with just a few moments' work, one can construct an analytic solution whose graph, to the naked eye, is indistinguishable from that of the exact solution, and whose relevant "parameter of quality" deviates from that of the exact result by less than 2%. The entire GRC procedure can be repeated, with the expected (but as yet unproven) result of decreasing even further, in a global way, errors that remain after the first GRC is employed.

For simplicity, consider the conservative Duffing equation, without damping or sources,

$$\frac{d^2x}{dt^2} = \ddot{x} = -x \cdot \phi(x), \qquad \phi(x) = x^2 - 1, \qquad (7.20)$$

a DE which has a long history and many current applications; almost all the statements made below will be relevant to more complicated, nonlinear $\phi(x)$. The first (energy) integral of this system is

$$E(D) = \frac{1}{2}\dot{x}^2 + V(x) = \frac{D^2}{2}\left(\frac{D^2}{2} - 1\right),$$

where $V'(x) = x \cdot \phi(x)$, and D is a maximum of $|x(t)|$, appearing here through the initial conditions: $x(0) = D$, $dx(0)/dt = 0$.

In the reference in Note 20, a "wavelet"-type of IR approximation was used to define an IR approximation to the Green's function of this problem, and methods of choosing sequential corrections to this IR approximation are suggested. However, it was noted there that by far the most efficient route to a simple, analytic approximation to the exact Jacobi elliptic functions (for $E(D)$ positive or negative) or to the exact solution for $E = 0$ ($x(t) = \sqrt{2}/\cosh(t)$) is obtained by selecting any reasonable IR "smoothed" solution, and rescaling that particular solution. We shall here choose the simplest possible form for a smoothed, IR approximation to the exact solution for $E(D) = 0$, and show how the RC procedure may be improved by using various forms of rescaling, and one in particular that takes into account time-reversal-invariance of the expected solution.

For such a second-order ODE, there are regions where one expects a positive $\ddot{x}/x$, and regions where this ratio is negative. From the DE, one sees that an inflection point may occur when $|x| = 1$, and we shall call the t-value where this takes place t_i, and the difference between t_i and the t-value at which our IR model solutions satisfy $|x| = 1$ will be taken as the measure of the quality of our approximation. From the exact Duffing calculation for $E(D) = 0$, and with the initial condition that $D = \sqrt{2}$ at $t = 0$, one sees that the solution begins in an "oscillatory" region, that is, with $\dot{x}/x$ negative; and that after x decreases in magnitude to values $x < 1$, it finds itself in an "exponential" region. Since $E(D)$ is fixed at 0, both $|\dot{x}|$ and x must continually decrease, as indicated by the DE, which shows that the asymptotic fall-off of $x(t)$ must take the form $\sim \exp[-t]$. This problem is simple because there is but one transition, from oscillatory (O) to an exponential (E) region; but the techniques can also be used[20] even when there are transitions back and forth between O and E zones.

As a "smoothed" IR approximation to the solution in the O region, we choose $x_O(t) = \sqrt{2}\cos(Mt)$, where M is a constant to be determined. If $x(t_1) = 1$, this requires $Mt_1 = \pi/4$, and substituting this into the expression for $E(D) = 0$ requires that $M = (2)^{-1/2}$, so that $t_1 = (\pi/4)\sqrt{2} \simeq 1.11$. In comparison, the exact inflection point occurs at $t_i = 0.881$, so that there is a fractional error of this relevant parameter of the amount $(t_1 - t_i)/t_i = 0.23$. In the E region, the solution must change to a decaying exponential, which we write as $x_E(t) = \exp[-N(t - t_1)]$, and using continuity of x and $\dot{x}$ at t_1, rapidly determine that $N = M = 1/\sqrt{2}$. One sees immediately, as plotted in Fig. 7.6, that the curves

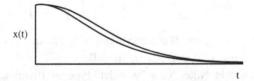

Fig. 7.6 A superposition of the simplest IR solution (top curve) with the exact, zero-energy solution (bottom curve).

of the exact and IR-approximated solutions are similar, but that the fall-off of the latter is too weak, compared to the former, for essentially all t.

We now invoke the GRC, which will bring the two solutions into very close agreement, for all t. The idea is to introduce a stretching, or a rescaling of all values of time, in both the O and E regions. In the former, we keep the form of the smoothed IR solution, but replace t by $t \cdot b(t)$, where $b(t)$ is a function to be determined,

$$x_0(t) = \sqrt{2}\cos(Mt \cdot b(t)),$$

and we expect that the O-to-E transition will now take place at a smaller time, t_2, where $t_2 < t_1$. In order to retain the same forms entering into the x and $\dot{x}$ continuity conditions at $x = 1$, one requires

$$[t \cdot b(t)]_{t_2} = t_1 \tag{7.21}$$

and

$$\frac{d}{dt}[t \cdot b(t)]_{t_2} = 1, \tag{7.22}$$

conditions which can be satisfied in a variety of ways. Following our "glissando" idea, we use (7.21) to rewrite (7.22) in the form $1 = b(t_2) + t_1 b'(t_2)/b(t_2)$, and accomplish this by finding a solution of the DE for variable $t < t_2$,

$$1 = b(t) + t_1 \cdot \frac{b'(t)}{b(t)}, \tag{7.23}$$

which is

$$b(t) = [1 - c\,e^{-t/t_1}]^{-1}. \tag{7.24}$$

Here, c is a constant to be determined by substituting the rescaled solution into the original DE, which identifies $b(0) = \sqrt{2}$, and hence $c = 0.293$. Finally, we determine t_2 by substituting (7.24) into (7.21), which yields $t_2 = 0.975$, to be compared with the previous value of $t_1 = 1.11$. Instead of an error of more than 20% in the relevant parameter, the rescaled solution has reduced the error to less than 7%.

With this new value of t_2, it is not necessary to resort to a DE to determine the scaling function in the E region, There, one already knows that the large-t fall-off is incorrect, and one may trivially use the rescaling method to correct that error. One simply writes the rescaled E-solution as $x_E(t) = \exp[-Nzb_E(z)]$, where $z = t - t_2$, and uses the conditions for no change of normalization of the continuity equations at $z = 0$ to obtain the pair of relations

$$[zb_E(z)]_{z=0} = 0, \quad \text{and} \quad \frac{d}{dz}[zb_E(z)]|_{z=0} = 1. \tag{7.25}$$

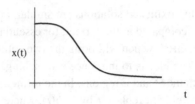

$x(t)$

t

Fig. 7.7 A superposition of the once-rescaled IR approximation to the exact solution.

A simple choice of function $b_E(z)$ which satisfies (7.25) and contains the correct asymptotic form is just the fraction

$$b_E(z) = (1 + \sqrt{2}z)/(1 + z), \qquad (7.26)$$

and one can then appreciate, from the superpositions of Fig. 7.7, the closeness of the new, GRC approximation to the exact result.

Can these rescaling operations be repeated to generate an even closer approximation to the exact answer? There is no reason why this cannot be done, except that the amount of work involved tends to become tedious. That is, in the O region, one can introduce a new variable $T = tb(t)$, and look for a solution $x_0(t) = \sqrt{2}\cos(MTB(T))$, where $B(T)$ is a rescaling function to be determined so that the new transition time, t_3, is even closer to t_i than was t_2. How fast such iterations converge – assuming they do – remains to be determined; but on the basis of the example of this section, one would expect a reasonable convergence.

One very definite method of increasing the accuracy of these rescaling corrections is to include any natural symmetry of the exact DE as part of the GRC procedure. In this case, one can evoke a simple "time-reversal-invariance", which consists of the replacement: $b(t) \rightarrow b(t^2)$. If we write the rescaled O region solution as

$$x_0(t) = \cos[(Mtb(t^2)],$$

then the scaling requirements analogous to (7.21) and (7.22) are given by

$$t_2 b(t_2^2) = t_1, 1 = [b + 2t^2 b']_{t=t_2} = [b + 2t_2 t_1 b'/b]_{t=t_2}, \qquad (7.27)$$

with $b'(u) = db/du$. If we choose the DE for $b(u)$ as

$$1 = b + 2t_1 t_2 \cdot \frac{b'}{b}, \qquad (7.28)$$

with solution

$$b(u) = [1 - c\,e^{-u/2t_1 t_2}]^{-1}, \tag{7.29}$$

where $b(0) = \sqrt{2}$ and $c = 0.293$, as before, one now finds that $t_2 = 0.893$, which is a very good approximation, corresponding to a relative error of the inflection point of 1.5%.

In summary, this "glissando" method of using a rescaling to insert a spectrum of frequencies higher than that, or those, chosen to provide a first IR approximation appears to work in a simple and straightforward fashion, and can provide quality approximations for nonlinear, integrable ODEs. What is not known are the convergence properties for multiple rescalings; nor have the complete set of such rescaling methods been enumerated and compared. But for situations where one desires an analytic, if approximate, solution for a given, nonlinear ODE, the method outlined above would seem to provide the necessary tools; and for this reason, the method would seem to justify serious mathematical study.

Notes

1 Using the same, functional notation as in the present volume, this reference is to the author's previous two volumes, here called HMF#1 and HMF#2.
2 The descriptions of wide-angle damping given in this chapter follow the underlying intuition (one hard- plus many soft-virtual exchanges) suggested by L. I. Schiff, *Phys. Rev.* **103** (1956) 443. Other authors have approached the same experimental results in different ways, e.g., R. Blankenbecler and R. L. Sugar, *Phys. Rev. D* **2** (1970), and other references quoted in HMF#2.
3 F. Bloch and A. Nordsieck, *Phys. Rev.* **52** (1937) 54.
4 D. R. Yennie, S. Frautschi, and H. Suura, *Ann. Phys. (NY)* **13** (1961) 379.
5 D. R. Yennie and H. Suura, *Phys. Rev.* **105** (1957) 1378.
6 J. Schwinger, *J. Math. Phys.* **2** (1961) 407.
7 K. Mahanthappa, *Phys. Rev.* **126** (1962) 329.
8 A. V. Svidinski, *JETP* **4** (1957) 179; and K. Symanzik, UCLA Lectures, 1960.
9 H. M. Fried and T. Gaisser, *Phys. Rev.* **179** (1969) 1491.
10 R. Jackiw, *Ann. Phys. (NY)* **48** (1968) 292.
11 T. T. Wu and C. N. Yang, *Phys. Rev.* **137** (1965) B708; T. T. Chou and C. N. Yang, *Phys. Rev. Letters* **20** (1968) 1213.
12 C. K. Akerlof, *et al.*, *Phys. Rev.* **159** (1967) 1138; H. M. Fried and T. K. Gaisser, *Phys. Rev. D* **4** (1971) 3330.
13 G. Källén, *K. Danske Vidensk. Selsk., mat-fys. Medd.* **27**, No.12 (1953).
14 M. Baker and K. Johnson, *Phys. Rev.* **183** (1969) 1292.
15 H. Cheng and T. T. Wu, *Expanding Protons: Scattering at High Energies*, MIT Press, Cambridge, MA (1987).
16 The needed Green's function is closely related to that of the "scalar laser" problem, Section 4.2.

17 A description of why and when eikonalization of simple exchanges is valid has been given by E. Eichten and R. Jackiw, *Phys. Rev. D* **4** (1971) 439.
18 For details, see the books of Notes 1 and 15.
19 B. Hasslacher and D. K. Sinclair, *Phys. Rev. D* **3** (1971) 1770; S.-J. Chang and T.-M. Yan, *Phys. Rev. D* **4** (1971) 537.
20 J.-D. Fournier and H. M. Fried, *Phys. Rev. D* **43** (1991) 3771. Some details of this section were worked out in an unpublished report by the same authors.

8

Models of high-energy, non-Abelian scattering

Ordered exponentials appear in Green's functions in a natural and fundamental way. There are, however, certain limiting situations in an Abelian theory where kinematic simplifications make it possible to avoid treating the OEs which happen to be present. (An example in QED is the neglect of $\sigma \cdot F$ dependence in the representation (3.31), and for two reasons: when the virtual NVMs exchanged are soft, each $F_{\mu\nu}$ is proportional to a small momentum; and, for small momentum transfers, the expectation value of any such spin quantity between the fermions' Dirac spinors will be very small. This is NOT true for wide-angle scattering in QCD, and the OEs which remain can give rise to interesting spin-polarization effects.)

What happens when the theory is fundamentally non-Abelian, when transfers of charge, or isospin, or of color degrees of freedom are to be expected between scattering particles? Can kinematic simplifications be found for the non-Abelian structure? For close-to-forward scattering in QCD, the QED quantities A_μ and $F_{\mu\nu}$ of (3.31) are to be replaced by $A^a_\mu \lambda_a$ and $F^a_{\mu\nu} \lambda_a$, where the λ_a are matrices of the fundamental, or defining representation of $SU(N)$. (For $N = 2$, $\lambda_a \to \sigma_a/2$, the Pauli matrices; for $N = 3$, the λ_a can be chosen as the Gell–Mann matrices.) The entire exponential dependence of such quantities must now be ordered, as in Chapter 3; and previous linkage operations upon the A_μ dependence can no longer be performed non-perturbatively, because the latter appear within the confines of a non-trivial OE.

In fact, there is a way to extract such A_μ dependence, so that the desired linkage operations can be explicitly performed, but it requires the introduction of an additional pair of FIs. In certain cases of limiting kinematics, such as scattering at very high energies, these new FIs can collapse into a set of ordinary quadratures, and – with appropriate care to maintain unitarity – may be used to advantage; these topics form the first two sections of this chapter. The remaining sections will return to Abelian physics, with the introduction of a special Green's function most useful for a Modified Multiperipheral Model (MMM), including all crossed and ladder graphs, in the tower-graph approximation to the eikonal scattering amplitude; and this will be followed by an explicit summation over all the eikonal graphs of the MMM. That result will then be "converted" to a model QCD result, by the artificial introduction of asymptotic freedom into the final answer.

8.1 An Abelian separation

Consider first a theory of fermions and bosons with an interaction Lagrangian density of the form $\mathcal{L}' = -g\bar{\psi}\gamma_\mu A^a_\mu \lambda_a \psi$, so that the relevant representation for an MSA $G_c[A]$ to be used for close-to-forward scattering will contain the OE

$$\left(\exp\left[ig \int_{s_B}^{s_A} ds' \lambda_a \frac{dQ(s')}{ds'} \cdot A^a(z - Q(s')) \right] \right)_+, \tag{8.1}$$

where the $Q(s')$ of (8.1) may refer to the quantity $\int_0^{s'} v$ of the original Fradkin representation (with $s_A \to s$, $s_B \to 0$); or, with different limits of integration, it may correspond to $2s'p$, as in the BN/no-recoil approximation. (Strictly speaking, for wide-angle scattering one should also retain the $\sigma \cdot F$ term as part of the OE; but, for simplicity, we ignore this complication here.)

Extraction of the A-dependence from the OE of (8.1) may be achieved by rewriting it in the form

$$N' \int d[\alpha] \int d[u] e^{i \int ds u_a(s) \alpha_a(s)} \left(e^{i \int ds \lambda_a u_a(s)} \right)_+ e^{-ig \int ds \alpha_a(s) \frac{dQ}{ds} \cdot A^a(z - Q(s))}, \tag{8.2}$$

where all the $\int ds$ integrations run over the same interval as that of (8.1), but the latter's prime of s', and the limits of integration, are suppressed. The normalization N' is such that the FI

$$N' \int d[\alpha] \exp\left\{ i \int \alpha_a(s) \left[u_a(s) - g A^a(z - Q(s)) \cdot \frac{dQ}{ds} \right] \right\}$$

is the delta-functional $\delta[u_a(s) - g \frac{dQ}{ds} \cdot A^a(z - Q(s))$, while subsequent $\int d[u]$ reproduces exactly the OE of (8.1). If the interval $\int ds$ is broken up into very small intervals labeled by s_i, of width Δs, where n factors of Δs equal the size of the integration range, then $N' = (N_i)^n$, where

$$N_i' = \left(\frac{\Delta s}{2\pi} \right)^D \tag{8.3}$$

and D is the number of dimensions over which each $\int d^p u(s_i)$ runs.

Of course, this separation does not solve the non-Abelian problem, but only postpones the evaluation of the OE of (8.2) until a later stage. The A-dependence, however, is now effectively Abelian, and operations upon it can be performed exactly. For example, consider the case of high-energy/small-momentum-transfer scattering of charged pions between nucleons, or gluons exchanged between quarks (neglecting, for the moment, all three- and four-gluon vertices, along with all closed-quark loops, so that the interactions between gluons, as they are exchanged between the scattering quarks, are suppressed). The linkage operator acting on the gluon vector potentials A_μ^a is then

$$\exp \mathcal{D}_{1,2} = \exp\left[-i \int \frac{\delta}{\delta A_{1,\mu}^{(a)}} D_{c,\mu\nu}^{ab} \frac{\delta}{\delta A_{2,\nu}^b} \right], \tag{8.4}$$

where $D_{c,\mu\nu}^{ab}(x-y)$ is the bare gluon propagator, which – for simplicity – we take as $\delta_{\mu\nu}\delta_{ab}D_c(x-y)$, with $\tilde{D}_c(k) = (\mu^2 + k^2 - i\epsilon)^{-1}$. The generic expression for the eikonal of this amplitude may then be represented as

$$e^{i\chi} = e^{\mathcal{D}_{1,2}} \left(e^{ig \int_{-\infty}^{+\infty} ds_1 \, p_1 \cdot A_1^a(z_1 - s_1 p_1)\lambda_a^{(1)}} \right)_+ \left(e^{ig \int_{-\infty}^{+\infty} dt_2 \, p_2 \cdot A_2^b(z_2 - t_2 p_2)\lambda_b^{(2)}} \right)_+ \Big|_{A_{1,2}\to 0}, \tag{8.5}$$

where $z_{1,2}$ and $p_{1,2}$ are the incident configuration-space and 4-momentum coordinates of the scattering quarks, respectively. The eikonal scattering amplitude corresponding to these graphs is then given by

$$T(s,t) = \frac{is}{2m^2} \int d^2b \, e^{iq_\perp \cdot b} \left[1 - e^{i\chi(s,b)} \right], \tag{8.6}$$

where $s = -(p_1 + p_2)^2$, $t = -(p_1 - p_1')^2$, m is the mass of each quark, and the impact parameter $\mathbf{b} = (\mathbf{z}_1 - \mathbf{z}_2)_\perp$ is the transverse (perpendicular to the incoming quark momenta) separation of the two quarks.

It should be emphasized that an integration over coupling constants, here suppressed, must really be performed at an earlier stage, before one can identify the RHS of (8.5) as $\exp[i\chi]$; but, for our purposes, (8.5) does express the qualitatively-correct form of that functional operation which can be identified as the eikonal of this problem. It should also be noted that there is much less structure in (8.5) than in the MPM of (7.19), because we have here dropped all linkages between the gluons that are exchanged between the quarks; in fact, (7.19) will be used as a model for the calculation of the latters' effects. The principal thrust of the present remarks is to find a way of performing the needed functional linkage operation when the original A-dependence is sitting inside an OE; and for this discussion it is simplest to neglect (temporarily) the interactions between the virtual gluons.

We now invoke the representation of (8.2), for each OE, and write

$$e^{i\chi} = N' \int d[\alpha] \int d[u] e^{i\int \alpha \cdot u} \left(e^{i\int \lambda^{(1)} \cdot u} \right)_+$$

$$\cdot N' \int d[\beta] \int d[v] e^{i\int \beta \cdot v} \left(e^{i\int \lambda^{(2)} \cdot v} \right)_+$$

$$\cdot e^{\mathcal{D}_{1,2}} e^{-ip_{1\mu} \int_{-\infty}^{+\infty} ds_1 \alpha_a(s_1) A_{1,\mu}^a(z_1 - s_1 p_1)} e^{-ip_{2\nu} \int_{-\infty}^{+\infty} dt_2 \beta_b(t_2) A_{2,\nu}^b(z_2 - t_2 p_2)} \Big|_0, \quad (8.7)$$

in which we may immediately perform the linkage operation of (8.5),

$$e^{\mathcal{D}_{1,2}} e^{-i\int \alpha \cdot p_1 \cdot A_1} \cdot e^{-i\int \beta p_2 \cdot A_2} |_0 = \exp\left[i\int \int_{-\infty}^{+\infty} ds \, dt \alpha_a(s) Q_{ab}(s,t) \beta_b(t) \right],$$

$$(8.8)$$

where $z_1 - z_2 = z$, $Q_{ab}(s,t) = g^2 \delta_{ab}(p_1 \cdot p_2) D_c(z - sp_1 + tp_2)$, and the dummy s, t variables are not to be confused with the Mandelstam scattering invariants. Perturbative, and then non-perturbative attempts to approximate the eikonal of (8.7) at extremely large energies, lead in a most natural way to the QAL of the next section.

8.2 The quasi-Abelian limit

There are several methods of approach to this problem – and to any non-Abelian scattering problem, at small or large momentum transfers – which lead to the idea that the Quasi-Abelian Limit (QAL)[1] is intuitively reasonable at ultra-high energies. Consider first the exponential term of the RHS of (8.8), and note that were this an Abelian problem, the α_a, β_b factors multiplying the NVM propagator D_c would be replaced by unity, as would the remaining FIs of (8.7), and the result would be the Abelian eikonal of (7.14). For this non-Abelian problem, in the limit of large energy E and small momentum transfer, (8.8) contains the CM 4-momenta $p_1 = (0,0,E;iE)$ and $p_2 = (0,0,-E;iE)$; and introduce there the following, rescaled, proper-time variables: $\bar{s} = Es, \bar{t} = Et$. The exponential factor of (8.8) then becomes

$$ig^2 \int_{-\infty}^{+\infty} d\bar{s} \int_{-\infty}^{+\infty} d\bar{t} \left(\frac{p_{1\mu}}{E} \right) \left(\frac{p_{2\nu}}{E} \right) \alpha_a \left(\frac{\bar{s}}{E} \right)$$

$$\cdot D_{c,\mu\nu}^{ab} \left(z - \bar{s} \left(\frac{p_1}{E} \right) + \bar{t} \left(\frac{p_2}{E} \right) \right) \beta_b \left(\frac{\bar{t}}{E} \right). \quad (8.9)$$

The ratios p_1/E and p_2/E are independent of E; and the only visible, overt energy dependence of (8.9) is that of the arguments of α_a and β_b. Imagine that a calculation is now carried out using (8.9), and that at the end of that calculation the limit $E \rightarrow \infty$ is taken. The leading results of that calculation might be expected to correspond precisely to what is obtained by taking the limit $E \to \infty$ under the s, t integrals in the relations which correspond to (8.9). Here, that limit would correspond to the replacement of (8.9) by

$$\mathrm{i}g^2 \alpha_a(0)\beta_b(0)\frac{p_{1\mu}}{E}\frac{p_{2\nu}}{E} \cdot \int\!\!\int_{-\infty}^{+\infty} \mathrm{d}\bar{s}\,\mathrm{d}\bar{t}\,D_{\mathrm{c},\mu\nu}^{ab}\left(z - \bar{s}\frac{p_1}{E} + \bar{t}\frac{p_2}{E}\right),$$

or, by what is the same (rescaled) thing:

$$\mathrm{i}g^2 \alpha_a(0)\beta_b(0)p_{1\mu}p_{2\nu} \int\!\!\int_{-\infty}^{+\infty} \mathrm{d}s\,\mathrm{d}t\,D_{\mathrm{c},\mu\nu}^{ab}(z - sp_1 + tp_2), \qquad (8.10)$$

in the limit of extremely large E. Were the α_a and β_b smooth, analytic functions of their arguments, one could argue that corrections to this limit would vanish as $E \to \infty$; but, although written in continuous form, the $\alpha_a(s)$ and $\beta_b(t)$ represent functions which are at best piecewise continuous, and no statements can be made about their derivatives. Yet, one has the intuitive expectation that the large E limit of the complete FIs without approximation – if such could be performed – would be very close to that generated by the QAL.

This is perhaps the simplest definition of the QAL, where one imagines that the non-perturbative result at high energy is correctly described by the great simplifications that follow from (8.10); two other approaches, which have the same QAL consequence, are described below. Mathematically, until one learns how to estimate corrections to this limit, the procedure is surely an unjustified interchange of limiting operations, for one is supposed to calculate all the FIs before allowing E to become arbitrarily large. Physically, this limit interchange suggests that sums and averages over all parameters of color exchange will, at very high energies, behave in the same way, and need be calculated just once – at $s = t = 0$ – because not enough proper time is available for fluctuations in the possible methods of color transfer; one might say[2] that the sum of all "color moments" effectively vanishes as $E \to \infty$. That is, regardless of the space–time point along a quark's trajectory where a virtual gluon is emitted or absorbed, the variables describing that color exchange are those associated with the quarks' distance of closest approach.

If this interchange of limits defining the QAL is reasonable, one sees that the only values of s and t which can enter into the non-trivial FIs over α_a, β_b, u_a, v_b are those of $s = t = 0$. Breaking up these integrals into distinct integrations over averaged variables carrying the values s_i and t_j, integration over all the $s_i \neq 0 \neq t_j$ intervals gives in each case, after extracting the proper parts of the normalization factors N', precisely a factor of unity. Each OE is then replaced by an un-ordered exponential factor depending on either $u_a(0)$ or $v_b(0)$, e.g.,

$$\left(e^{i\int_{-\infty}^{+\infty} ds\lambda_a u_a(s)}\right)_+ = e^{i\Delta s\lambda \cdot u(s_n)} \cdots e^{i\Delta s\lambda \cdot u(s_1)} \cdot e^{i\Delta s\lambda \cdot u(0)}$$

$$\cdot e^{i\Delta s\lambda \cdot u(s_{-1})} \cdots e^{i\Delta s\lambda \cdot u(s_{-n})}\Big|_{\substack{s_n > \cdots > s_1 > 0 > s_{-1} > \cdots > s_{-n}, \\ s_n \to +\infty, s_{-n} \to -\infty}},$$

which gives just

$$e^{i\Delta s\lambda \cdot u(0)},$$

because the integrals $\int d^D \alpha(s_i)$ and $\int d^D \beta(t_j)$ for s_i and t_j produce factors of $\delta(u_a(s_i))$ and $\delta(v_b(t_j))$, so that each OE reduces to the un-ordered form above.

For $SU(2)$, for simplicity, where the D of (8.3) equals 3, what remains is the set of quadratures

$$e^{i\chi} = \left(\frac{\Delta s}{2\pi}\right)^6 \int d^3\alpha(0) \int d^3\beta(0) \int d^3u(0) \int d^3v(0) \cdot e^{i\Delta s[\alpha(0)\cdot \mathbf{u}(0) + \beta(0)\cdot \mathbf{v}(0)]}$$

$$\cdot e^{\frac{i}{2}\sigma_a^I \Delta s u_a(0)} \cdot e^{\frac{i}{2}\sigma_b^{II} \Delta s v_b(0)} \cdot e^{-iK\alpha_a(0)\beta_a(0)}, \tag{8.11}$$

where $K = (g^2/2\pi)K_0(Mb)$, and for simplicity, we have chosen $D_{c,\mu\nu} = \delta_{ab}\delta_{\mu\nu}\Delta_c$. A trivial change of variables, $\Delta s u_a(0) = u_a$, $\Delta s v_b(0) = v_b$, $\alpha_a(0) = \alpha_a$, $\beta_b(0) = \beta_b$, converts (8.11) into

$$e^{i\chi} = (2\pi)^{-6} \int d^3\alpha \int d^3\beta \int d^3u \int d^3v \exp[i(\alpha \cdot \mathbf{u} + \beta \cdot \mathbf{v} - i\alpha \cdot \beta K)].$$

$$\cdot \exp\left[\frac{i}{2}\sigma^I \cdot u\right] \cdot \exp\left[\frac{i}{2}\sigma^{II} \cdot v\right]. \tag{8.12}$$

Integration over $\int d^3\alpha$, $\int d^3\beta$ is easily performed, yielding, after another rescaling,

$$e^{i\chi} = (2\pi)^{-3} \int d^3u \int d^3v\, e^{i(\mathbf{u}\cdot\mathbf{v})} \cdot e^{\frac{1}{2}\sigma^I \cdot u\sqrt{K}} \cdot e^{\frac{1}{2}\sigma^{II} \cdot v\sqrt{K}}. \tag{8.13}$$

Were the $\sigma_i^{I,II}$ of (8.13) treated as ordinary numbers, the remaining integrals would immediately generate what might be called the naive result

$$e^{i\chi_0} = e^{-i(\sigma^I \cdot \sigma^{II})K/4}; \tag{8.14}$$

however, a more careful, if elementary, evaluation of (8.14) is needed, which yields

$$e^{i\chi} = \cos\left(\frac{K}{4}\right) - \left(\frac{K}{4}\right)\sin\left(\frac{K}{4}\right) - \frac{i}{3}(\sigma^I \cdot \sigma^{II})$$

$$\cdot \left[\sin\left(\frac{K}{4}\right) + \left(\frac{K}{4}\right)\cos\left(\frac{K}{4}\right)\right]. \tag{8.15}$$

Since the product $\sigma^{\mathrm{I}} \cdot \sigma^{\mathrm{II}}$ has eigenvalues of $+1$ (triplet state) and -3 (singlet state), the singlet eikonal function is then given by

$$e^{i\chi s} = \left(1 + i\frac{K}{4}\right)e^{iK/4} = \rho e^{i\left(\frac{K}{4} - \theta\right)}, \qquad (8.16)$$

where $\rho = [1 + (\frac{K}{4})^2]^{1/2}$ and $\theta = \tan^{-1}(K/4)$. Using the same notation, the triplet eikonal may be written as

$$e^{i\chi_T} = \frac{\rho}{3}e^{i\left(\frac{K}{4} - \theta\right)} + \frac{2\rho}{3}e^{-i\left(\frac{K}{4} - \theta\right)}. \qquad (8.17)$$

Assuming the validity of the QAL, these are the correct, high-energy eikonals of this problem, which is equivalent to that of nucleon-nucleon scattering by the exchange of charged rho-mesons.[3]

Generalizations from $SU(2)$ to $SU(3)$ are possible, although somewhat tedious; here, for any $SU(N)$, a convenient representation for needed exponentials can be written in the form

$$e^{i\boldsymbol{\lambda}\cdot\mathbf{u}} = \frac{1}{N}\sum_n \left[1 + \lambda_a \frac{\partial r_n}{\partial u_a}\right]e^{ir_n}, \qquad (8.18)$$

where the r_n are the eigenvalues of the matrix $\boldsymbol{\lambda}\cdot\mathbf{u}$. For $SU(3)$, for example, one must solve the triplet of equations

$$\sum_n r_n = 0, \quad \sum_n r_n^2 = \bar{a}\cdot\delta_{ab}(u_a u_b), \quad \sum_n r_n^3 = \bar{b}(d_{abc}u_a u_b u_c),$$

where $\bar{a}$ and $\bar{b}$ are real constants, for the three r_n, which is equivalent to finding the roots of a relevant cubic equation; for $SU(2)$, one has, immediately, $e^{\frac{i}{2}\sigma\cdot\mathbf{u}} = \cos(\frac{u}{2}) + i\frac{\sigma\cdot\mathbf{u}}{u}\sin(\frac{u}{2})$.

We next turn to a partial justification of the QAL, and attempt to do this with two distinct arguments, as follows.

(1) Imagine a perturbative calculation of the eikonal, in which there appear multiple commutators, of the form

$$[\lambda_{a_1}, [\lambda_{a_2}, \cdots [\lambda_{a_{n-1}}, \lambda_{a_n}]\cdots]]$$

corresponding to variations in the color (for $SU(3)$) or isotopic (for $SU(2)$) coordinates along each fermion line. Perturbatively, these commutators – and here we do *not* include perturbative contributions coming from other, interior parts of the graphs – generate multiple $\ln(E/m)$ factors; and it is possible to extract the leading-log dependence of every such nth perturbative order, which appears in the form $(x)^n$, where $x = K\ln(E/m)$, and $K = (g^2/2\pi)K_0(Mb)$. In order to sum such contributions over all $n > 0$, one must limit the magnitude of x by $|x| < 1$, and the result is: $x/(1-x)$. If one now takes the limit as $x \to \infty$, one sees that all the leading-log terms, in effect, self-cancel, and produce for each quark line a factor of -1. The simple and succinct way of bypassing all the effort of this example is to adopt the QAL, since the latter procedure will display *no* $\ln(E/m)$ dependence at all for these contributions, neither leading-log, nor next-to-leading-log, etc.

One can see that this is just the content of the QAL, by noting that the product of two relevant exponential factors is not equal to the product of their sum,

$$e^{i\lambda \cdot a} e^{i\lambda \cdot b} \neq e^{i\lambda \cdot (a+b)}.$$

But if we believe that the sum of such perturbative, leading-log terms is going to self-cancel, let us neglect all such multiple commutator dependence, and calculate the result. For this, it is useful to note that the FIs over the α_i and β_j of (8.7), with (8.8), can be performed exactly, with the result that the remaining FIs of (8.7) may be replaced by the (asymmetric but convenient) form

$$\prod_{i,j} (2\pi)^{-D} \int d_{u_i}^D \int d_{v_j}^D e^{-i u_i \cdot v_j} \cdots e^{+i\lambda^{\mathrm{I}} \cdot \sum_k \Delta_{c,ik} u_k} \cdots e^{i\lambda^{\mathrm{II}} \cdot v_j} \cdots, \qquad (8.19)$$

where the $\Delta_{c,ik}$ denote discrete s_i, t_k values of $g^2(p_1 \cdot p_2)\Delta s \Delta t D_c$, and where the dots in (8.19) indicate the exponential λ^{I}, λ^{II} factors with all the differing values of s_i and t_j. If we now neglect all such commutator dependence, we may replace in (8.19)

$$\prod_i e^{i\lambda^{\mathrm{I}} \cdot \sum_k \Delta_{c,ik} u_k} \quad \text{by} \quad e^{i\lambda^{\mathrm{I}} \cdot \sum_{i,k} \Delta_{c,ik} u_k},$$

or in the continuum limit by

$$\exp\left[ig^2(p_1 \cdot p_2) \int\!\!\int_{-\infty}^{+\infty} ds\, dt\, \Delta_c(z - sp_1 + tp_2) u(t) \cdot \lambda^{\mathrm{I}}.\right.$$

Inserting the Fourier representation for Δ_c and integrating over $\int ds$ then generates a $\delta(k_3 - k_0)$, which permits the $\int dk_0$ to be performed, and generates the simple result $\exp[-iK\lambda^{\mathrm{I}} \cdot u(t_0)]$, where $t_0 = (z_3 - z_0)/2E$. For large E, we may take t_0 to be equal to zero (the classical coordinates for $z = z_1 - z_2$ may be defined so that the differences $z_3 - z_0$ in the CM will cancel, regardless of the value of E), and the result (in $SU(2)$) will be just a rescaling away from that of (8.13). In other words, the self-cancelling $\ln(E/m)$ dependence coming from the multiple commutators along each fermion line may be thought of as the essence of the QAL.

(2) A second method of approach to the QAL utilizes the fall-off (in space-like regions, or rapid oscillations in time-like regions) behavior of the propagator

$\Delta_c(Z)$, which decreases roughly as $\exp[-M(Z^2)^{1/2}]$ for large values of $Z^2 = (z - sp_1 + tp_2)^2$. In our case, for s and $t \neq 0$, this means a propagator fall-off as roughly $\exp[-MEs_it_j]$. In terms of the s, t variables, the region of importance is within the "star" formed by the hyperbolae $|s_it_j| < (ME)^{-2}$; all other s, t will give a negligible contribution to the product $\alpha(s_i)\Delta_{c,ij}\beta(t_j)$. It should be noted that arbitrarily large α_i, β_j values are allowed, but then the entire interaction exponential oscillates to zero. The point of the exercise is that, in the limit $E \to \infty$, the "star" shrinks to the point $s \sim t \sim 0$, so that, again, only these are the significant values.

With these mathematical "justifications", and with the physically intuitive picture of color transfers between scattering fermions taking place at high energies very close to their distance of closest approach, one has in QAL an approximate tool at non-asymptotic energies which one can employ, in a non-perturbative way, for calculations relevant to high-energy scattering. For example, it should be possible to use QAL in evaluating wide-angle, non-Abelian scattering, and in particular to obtain statements describing the observed, nucleon-spin-polarization experiments, which still demand non-perturbative explanation.[4] And it should be possible to extend the analysis of Section 8.4 to the high-energy elastic scattering of quarks, bound in distinct nucleons.

8.3 Loop, ladder, and crossed-ladder approximations

Prior to displaying a model Green's function of relevance to the question "What is the effect of all the neglected, non-tower eikonal graphs?", we first summarize the situation which has been known for several decades; and we shall emphasize possible forms, predictions, and experimental values for total cross sections, $\sigma_{\text{TOT}}(s)$ (rather than for the other quantities of interest, such as differential cross sections, and shrinkage of elastic-scattering diffraction peaks). Here, s will refer to the Mandelstam variable, corresponding to the square of the total CM energy of the incident particles.

In Chapter 7, we described three possible forms of eikonal, which result from different classes of graphs exchanged between a pair of scattering fermions. These are:

(a) The simplest eikonal of (7.14), obtained by the interchange of all possible virtual NVMs between the fermions; this real χ_1 produces a constant σ_{TOT} as s becomes arbitrarily large. Physically, and as noted in Chapter 7, this model's shortcoming is that it contains no mechanism for the inelastic production of large numbers of particles at very small momentum transfers.

(b) Towers of rungs of closed fermion loops, attached to each other and to the scattering fermions by a pair of NVMs, generate the original Cheng–Wu eikonal exhibited in (7.17). This eikonal function is imaginary, corresponding to a diminution of the elastic amplitude as inelastic production occurs, as required by unitarity. Such towers are able to model significant particle production at small momentum transfers, and generate a σ_{TOT} which increases as $\ln^2 s$, the form of the maximum such dependence allowed by the Froissart bound.

(c) Towers composed of ladder graphs, with rungs of scalar quanta, exchanged between a pair of NVMs, yield the eikonal of (7.18), which is qualitatively equivalent to that of (7.17), and for the same physical reason.

It will be useful to illustrate how the eikonals of (b) and (c) generate their $\sigma_{\text{TOT}}(s)$, and for this we shall consider the simplest case of all, that of the original Cheng–Wu eikonal, for which one may write $i\chi = -\rho(s, b)$, $\rho(s, b) = a \cdot s^{\alpha} \exp[-\mu b]$. Equation (7.15) may then be written as

$$\sigma_{\text{TOT}} = 2 \int d^2 b \left[1 - e^{-\rho(s,b)} \right], \qquad (8.20)$$

and to evaluate the form of the resulting s-dependence it is convenient to define a quantity $b_0(s)$ by the relation: $1 = \rho(s, b_0(s))$; that is, $b_0(s) \simeq \frac{\alpha}{\mu} \ln s + \cdots$ is that value of impact parameter where any increase of ρ with increasing s is just counterbalanced by the damping with respect to b. One sees from (8.20) that for $b < b_0$, ρ is large and the $\exp[-\rho]$ is small, so that this contribution gives essentially

$$2 \int_0^{b_0} d^2 b [1] \simeq 2\pi b_0^2(s) \sim \ln^2 s + \cdots . \qquad (8.21)$$

In contrast, for $b > b_0$, ρ is small, and the exponential of (8.20) may be expanded, and that portion of the integral approximated by

$$2 \int_{b_0}^{\infty} d^2 b \cdot \rho \sim \ln s + \cdots \qquad (8.22)$$

which is down by one factor of $\ln(s)$ compared to the leading s-dependence of σ_{TOT}, which arises from large $b \sim \ln(s)$. It is left as an exercise for the interested reader to repeat this calculation using the eikonal of case (c), and to verify that both (b) and (c) describe the same Physics, at least in the form of their predictions for $\sigma_{\text{TOT}}(s)$.

It should be emphasized that these ladder-graph calculations have made use of an additional, unjustified approximation, by retaining only the "leading-log" terms of every perturbative order; that is, if the coupling constant of the NVM to the scalar particle is G, a ladder graph with n rungs contributes an amount proportional to $G^n[\ln(s)]^n$, in contrast to a ladder in which one pair of rungs is crossed, of contribution proportional to $G^n[\ln(s)]^{n-1}$; every time another pair of rungs is crossed, the $\ln(s)$ dependence drops by another power. The total number of such graphs is given by $n!$, and if only those terms with the largest powers of $\ln(s)$ are kept, this means that $(n! - 1)$ "less-important" terms are discarded, an approximation that is mathematically untenable for n sufficiently large such that $n!$ is of the order of $\ln(s)$, or larger. Nevertheless, for reasons of "simplicity" – one calculates what one can, and hopes for the best – this type of approximation has long been made, without justification.

Another approximation, made for the same reason, has been to neglect contributions coming from eikonal graphs more complicated than the towers. To even attempt such a calculation one is forced into a functional description, for the number of classes of graphs which must be included, corresponding to the exchange of all possible t-channel NVMs, between which are exchanged all possible numbers of scalar mesons, is simply staggering. Such a functional description, given in (7.19), has been known for three decades; but what was lacking was a suitable NMV propagator $\bar{\Delta}_c(x, y|\pi)$ in a fictitious scalar field, which could realistically model the inelastic production of scalar particles, here called "scalar pions", and which could be used to estimate the elastic-scattering eikonal of (7.19). Such a Green's function is presented immediately below, and is used to suggest one possible form of the eikonal in a "generalized" Cheng–Wu context, containing towers formed from ladders and crossed-rung ladders, in all possible combinations; and to compare the result with the leading-log eikonals of cases (b) and (c). In the next section, the calculation is extended to include the sum of all eikonal graphs of this model, without exception.

As a preliminary step, the reader is referred to the functional cluster expansion of Chapter 2, in particular to

$$\exp\left[-\frac{i}{2}\int\frac{\delta}{\delta\pi}D_c\frac{\delta}{\delta\pi}\right]\cdot\exp[L[\pi]] = \exp\left[\sum_{N=1}^{\infty}\frac{Q_N}{N!}\right], \qquad (2.46)$$

where the Q_N are the connected cluster functionals, derived in subsequent paragraphs of that section. In the present context, the functional $L[\pi]$ is given by the RHS exponential factor of (7.19), and is operated upon by the linkage operator, as shown. Because we are interested only in the exchange of virtual πs *between* the virtual NVMs – that is where the Physics of this problem lies, governed by unitarity – we first drop all radiative corrections *along* each NVM, which is a simplification that can easily be performed functionally. In particular, the NVM mass M and coupling G to the πs are taken as "renormalized" constants (at least until the end of the next section, when a closer correspondence of this model with QCD is attempted).

Dropping all radiative corrections along the NVMs, the quantity Q_1, as defined in (2.47), becomes just the $i\chi_1$ of (7.14), while the Q_2 of (2.48) may be written as

$$Q_2 = \left[e^{-i\int \frac{\delta}{\delta\pi_a} D_c \frac{\delta}{\delta\pi_b}} - 1\right] L[\pi_a] L[\pi_b]|_{\pi_{a,b}\to 0}, \qquad (8.23)$$

and is the "tower graph" approximation to the eikonal of this problem, with all numbers of virtual πs exchanged, as ladders and crossed-rung ladders, between one pair of NVMs. The Q_N for higher powers of N correspond to corrections to this tower eikonal, constructed from π-exchange between more than two NVMs, and in all possible ways between these multiple NVMs.

The next step is the specification of a suitable Green's function, $\bar{\Delta}_c[\pi]$, which can model the emission of relatively high-energy πs from each NVM. Experimentally, most of the momenta of particles emitted inelastically are in close-to-forward directions, and so we may imagine that the field $\pi(x)$ depends only on x_3 and x_0, with transverse momentum components subsequently limited (which is also in agreement with experimental inelastic emissions) in another, model-dependent way. Because these emitted particles (gluonic jets, in QCD) are of high energy – or, more accurately, we wish to extract those parts of these individual processes which increase as $\ln(s)$ – we can assume that these are all relativistic particles, and replace $\pi(x_3, x_0)$ by $\pi(x_3 - x_0)$.

This form suggests particles moving relativistically in the $+x_3$ direction; but whether that direction lies in the $+z_3$ direction of the CM depends on how these x-variables connect to the pion propagator $D_c(u - v)$. The latter is perfectly relativistic, in the sense that it contains both particle and anti-particle poles in its k_0-plane; and it will generate a logarithmic divergence corresponding to particles emitted in the $\pm z_3$ directions, as dictated by energy–momentum considerations. That log divergence corresponds to one that would be found in the probability for emitting a scalar pion of arbitrarily high longitudinal momentum; and just as is done for the ordinary tower (ladder) graphs, we regulate that log divergence by the physically-sensible requirement that each $(k_3)_{max} \sim \sqrt{s}$. The model is completed by inserting, by hand, a $k_\perp$ cut-off in the definition of D_c, and one can then study the effects of doing this for different types of $k_\perp$ cut-offs.

These physically-motivated restrictions and insertions define the model, which can then be used to reproduce the essential results of the ladder-graph towers, and to explore the "internal, unitarity cancellations" which one might expect to result from summing over all the eikonal graphs, as in the next section.

However, even with the physical attributes of this model, one is not yet able to perform the full analysis without an additional assumption, which, eventually, must be justified.

The model's simplicity can now be realized by rewriting $\pi(x_3 - x_0)$ as $\pi(n^{(-)} \cdot x)$, and by recognizing the similarity of the Green's function with that of the "scalar laser" solution for $G_c[A]$ of Chapter 4, where $A(x)$ was written as $A(k \cdot x) = A(\omega n^{(-)} \cdot x)$, for $k_\mu = (0, 0, \omega; i\omega)$. All the results of that analysis may be taken over immediately by replacing k_μ of (4.16) by k_μ/ω, so that one can write $\bar{\Delta}_{c,\mu\nu} = \delta_{\mu\nu} \bar{\Delta}_c$, with

$$\bar{\Delta}_c(x, y|\pi) = \frac{1}{16\pi^2} \int_0^\infty \frac{ds_a}{s_a^2} e^{-is_a m^2 + i\frac{(x-y)^2}{4s_a}} \cdot e^{-iGs_a \int_0^1 d\lambda \pi(n^{(-)} \cdot \xi(x, y(\lambda))}, \quad (8.24)$$

where $\xi_\mu = \lambda x_\mu + (1 - \lambda) y_\mu$, and where we will use the subscripts $a, b, c, \ldots$ to distinguish the different NVM propagators.

Each such Green's function enters into (7.19) in the form

$$ig^2(p_1 \cdot p_2) \iint_{-\infty}^{+\infty} d\bar{s}\, d\bar{t}\, \delta(u - [z_1 - \bar{s} p_1])\delta(v - [z_2 - \bar{t} p_2])\bar{\Delta}_c(u, v|\pi)$$

$$= \frac{ig^2(p_1 \cdot p_2)}{16\pi^2} \int_0^\infty \frac{ds_a}{s_a^2} e^{-is_a m^2} \iint_{-\infty}^{+\infty} d\bar{s}_a\, d\bar{t}_a\, e^{\frac{i}{4s_a}(z_{12} - \bar{s}_a p_1 + \bar{t}_a p_2)^2}$$

$$\cdot e^{-iGs_a \int_0^1 d\lambda_a \pi(n^{(-)} \cdot \xi_a(\lambda_a))}$$

where $\xi_a(\lambda_a) = \xi_a(z_1 - \bar{s}_a p_1, z_2 - \bar{t}_a p_2|\lambda_a)$, and the "bar" notation does not imply a rescaling (as in the previous section). Restricting the calculation to Q_2, which will yield this model's version of $i\chi_2$, from (8.23) we need calculate

$$\left[\frac{ig^2(p_1 \cdot p_2)^2}{16\pi^2}\right]^2 \int_0^\infty \frac{ds_a}{s_a^2} \int_0^\infty \frac{ds_b}{s_b^2} e^{-im^2(s_a + s_b)}$$

$$\cdot \iint_{-\infty}^{+\infty} d\bar{s}_a\, d\bar{t}_a \iint_{-\infty}^{+\infty} d\bar{s}_b\, d\bar{t}_b \cdot e^{\frac{i}{4s_a}(z_{12} - \bar{s}_a p_1 + \bar{t}_a p_2)^2} \cdot e^{\frac{i}{4s_b}(z_{12} - \bar{s}_b p_1 + \bar{t}_b p_2)^2}$$

$$\cdot \left[e^{-i\int \frac{\delta}{\delta\pi_a} D_c \frac{\delta}{\delta\pi_b}} - 1\right] \cdot e^{-iGs_a \int_0^1 d\lambda_a \pi_a(n^{(-)} \cdot \xi_a(\lambda_a))}$$

$$\cdot e^{-iGs_b \int_0^1 d\lambda_b \pi_b(n^{(-)} \cdot \xi_b(\lambda_b))}\Big|_{\pi_{a,b} \to 0}, \quad (8.25)$$

where $\tilde{D}_c(k) = e^{-\gamma^2 k_\perp^2}[\mu^2 + k_3^2 - k_0^2 - i\epsilon]^{-1}$, and γ is the $k_\perp$ cut-off to be specified below.

The last lines of (8.25) are immediate,

$$e^{i\int_0^1 d\lambda_a \int_0^1 d\lambda_b \cdot G^2 s_a s_b D_c(\xi_a^{(-)}(\lambda_a) - \xi_b^{(-)}(\lambda_b))} - 1, \quad (8.26)$$

while the propagator of (8.26) may be written as

$$\int \frac{d^2k_\perp}{(2\pi)^2} \, e^{-\gamma^2 k_\perp^2} \cdot \frac{1}{2} \int\int \frac{dk_{(+)}\, dk_{(-)}}{(2\pi)^2} \, \frac{e^{\frac{1}{2}k_{(+)}(\xi_a^{(-)}(\lambda_a)-\xi_b^{(-)}(\lambda_b))}}{\mu^2 + k_{(+)}k_{(-)} - i\epsilon}, \qquad (8.27)$$

where $k_{(\pm)} = k_3 \pm k_0$. With $\mathcal{Z} = \frac{\xi_a^{(-)} - \xi_b^{(-)}}{2}$, the $k_{(\pm)}$ integrals of (8.27) may be rewritten as

$$\frac{1}{2} \cdot \frac{1}{(2\pi)^2} \cdot \int \frac{dk_{(-)}}{k_{(-)}} \int dk_{(+)} e^{ik_{(+)}\mathcal{Z}} \left[\frac{\mu^2}{k_{(-)}} + k_{(+)} - i\epsilon \cdot \epsilon\big(k_{(-)}\big) \right]^{-1},$$

where $\epsilon = 0+$, and $\epsilon(x) = \theta(x) - \theta(-x)$. Integration over $k_{(+)}$ depends on the sign of $k_{(-)}$ and yields

$$(2\pi i)\big\{ \theta\big(k_{(-)}e\big) e^{-i|\mathcal{Z}|\mu^2/k_{(-)}} - \theta\big(-k_{(-)}\big) e^{+i|\mathcal{Z}|\mu^2/k_{(-)}} \big\}, \qquad (8.28)$$

so that both terms of (8.28) contribute equally to the remaining $\int dk_{(-)}$ yielding

$$\frac{i}{2\pi} \int_0^\infty \frac{dk}{k} \, e^{-i|\mathcal{Z}|\mu^2/k}. \qquad (8.29)$$

This integral diverges logarithmically for large k, and as explained above, we insert a cut-off $k_{max} \sim \sqrt{s}$, and an arbitrary scale-parameter m, to obtain the dominant contribution for large s: $\frac{i}{4\pi} \ln(s/m^2)$. In this way, (8.27) becomes $\frac{i}{(4\pi\gamma)^2} \ln(s/m^2)$; and because this leading s-dependence is independent of $\lambda_{a,b}$, (8.26) simplifies to

$$\exp\left[-\alpha_G \left(\frac{s_a s_b m^2}{4\pi\gamma^2} \right) \ln(s/m^2) \right] - 1, \qquad (8.30)$$

where

$$\alpha_G = \frac{G^2}{m^2}/4\pi, \quad \text{and} \quad \alpha_g = g^2/4\pi.$$

The parametric integrals over $\bar{s}_{a,b}, \bar{t}_{a,b}$ still remain to be done, with (8.25) replaced by

$$-\frac{\alpha_g^2 s^2}{4(4\pi)^2} \int_0^\infty \frac{ds_a}{s_a^2} \int_0^\infty \frac{ds_b}{s_b^2} e^{-im^2(s_a+s_b)} \cdot \left[e^{-\frac{\alpha_G m^2}{4\pi\gamma^2} \cdot s_a s_b \cdot \ln(s/m^2)} - 1 \right]$$

$$\cdot \int\int_{-\infty}^{+\infty} d\bar{s}_a \, d\bar{t}_a \int\int_{-\infty}^{+\infty} d\bar{s}_b \, d\bar{t}_b \, \exp\left[\frac{i}{4s_a}(z_{12} - \bar{s}_a p_1 + \bar{t}_a p_2)^2 \right.$$

$$\left. + \frac{i}{4s_b}(z_{12} - \bar{s}_b p_1 + \bar{t}_b p_2)^2 \right],$$

and with $p_{1\mu} = En_\mu^{(-)}$, $p_{2\mu} = -En_\mu^{(+)}$, $\mathbf{b} = (\mathbf{z}_{12})_\perp$, those integrals display a lovely cancellation of all non-transverse z_{12} dependence, and generate

$$i\chi_2 = -\alpha_g^2 \int_0^\infty \frac{ds_a}{s_a} \int_0^\infty \frac{ds_b}{s_b} e^{-im^2(s_a+s_b)+\frac{ib^2}{4}(\frac{1}{s_a}+\frac{1}{s_b})} \cdot \left[e^{-m^2\alpha_G(\frac{s_a s_b}{4\pi\gamma^2})\ln(s/m^2)} - 1 \right]$$

or

$$i\chi_2 = -\alpha_g^2 \sum_{n=1}^\infty \frac{1}{n!} \left(\frac{m^2\alpha_G}{4\pi\gamma^2} \right)^n \ln^n(s/m^2) \int_0^\infty \frac{ds_a}{s_a}$$

$$\cdot \int_0^\infty \frac{ds_b}{s_b} (-s_a s_b)^n e^{-im^2(s_a+s_b)} \cdot e^{i\frac{b^2}{4}(\frac{1}{s_a}+\frac{1}{s_b})}. \tag{8.31}$$

It is now convenient to make the standard continuation: $s_a \to -i\tau_a$, $s_b \to -i\tau_b$, so that (8.31) becomes

$$i\chi_2 = -\alpha_g^2 \sum_{n=1}^\infty \frac{1}{n!} \left(\frac{m^2\alpha_G \ln(s/m^2)}{4\pi\gamma^2} \right)^n \left[\int_0^\infty \frac{d\tau}{\tau} \tau^n e^{-m^2\tau - b^2/4\tau} \right]^2. \tag{8.32}$$

Since the integral inside the squared bracket of (8.32) is proportional to the Bessel function $K_n(mb)$, an alternative expression is

$$i\chi_2 = -4\alpha_g^2 \sum_{n=1}^\infty \frac{1}{n!} \left[\frac{\alpha_G}{16\pi} \left(\frac{b^2}{\gamma^2} \right) \ln(s/m^2) \right]^n K_n^2(mb). \tag{8.33}$$

This eikonal is properly absorptive, but appears too complicated to be inserted into (7.15) and evaluated in a straightforward (and finite) manner. But since we expect a strong correlation between the behavior of $\sigma_{\text{TOT}}(s)$ and large b values [$b \sim b_0(s) \sim \ln(s)$], we can ask if (8.33) simplifies in the limit of large b; and this is exactly the case for $b \gg 1/m$, which is certainly expected, since one assumes that $\ln(s/m2) \gg 1$. Note that, for large impact parameter, the only natural $k_\perp$ cut-off in the model propagator is b itself, and this is our choice: $\gamma = b$.

One would like to be able to replace each $K_n(mb)$ of (8.33) by its large-mb asymptotic form, $[\frac{\pi}{2mb}]^{1/2} e^{-mb}[1 + \cdots]$, but any such interchange of sum and asymptotic limit must be viewed with suspicion, and justified, even if the result is so reasonable that it is not difficult to suppress disbelief. The mathematically-improper step that one would like to take, for $mb \gg 1$, is to replace each $K_n(mb)$ of (8.33) by its asymptotic form above, with leading term independent of n, so that the sum again exponentiates. This is correct only if $mb > (n^2 - 1)/2$, as is easily seen by examining the next terms of the expansion.[5] Were there only a finite number of such large-n correction terms, one could argue that the essential results of that approximation would be correct. But the sum of

(8.33) runs over all n; and hence this simplifying approximation is clearly incorrect.

In fact, what this does suggest is that the model needs to be refined so that a new set of functions $H_n(mb)$, whose asymptotic n-dependence is sufficiently weak, should replace the $K_n(mb)$ of (8.33), in order to permit the interchange of limits stated above. This can be accomplished if one imagines that the neglected self-energy structure along each $\bar{\Delta}_c[\pi]$ line is included, with a net effect of damping away large-τ contributions to the previous representations. The model used above for extracting the non-perturbative forms of high-momentum linkages between different $\bar{\Delta}_c[\pi]$ is not necessarily the one appropriate for the less-energetic self-linkages along each $\bar{\Delta}_c[\pi]$; but were it used for the latter, an extra factor of $\exp[-iG^2 \cdot \tau^2 \cdot I/2]$, $I = \int_0^1 d\lambda \int_0^1 d\lambda^1 D_c(\xi_{(\lambda)}^{(-)} - \xi^{(-)}(\lambda^1))$, would appear in each τ-integral, and for Re $I \neq 0$, and/or Im $I < 0$, would generate significant damping for very large τ.

A completely different, soluble model is one in which the original, Fradkin-variable statement of all possible self-linkages is exactly expressed by the exponential of $i\frac{G^2}{m^2} \int_0^s ds_1 \int_0^{s_1} ds_2 D_c(\int_{s_2}^{s_1} ds' v(s'))$, and where the latter quantity is then approximated in a "no-recoil" fashion by $i\frac{G^2}{m^2} \int_0^s ds_1 \int_0^{s_1} ds_2 D_c(v_0(s_1 - s_2))$, with v_0 corresponding to an "averaged" NVM 4-velocity, such that $v_0^2 = -1$. For zero-propagator mass, it is well-known that this propagator can be expressed exactly by

$$D_c(z) = \left(\frac{i}{4\pi^2}\right)\frac{1}{z^2 + i\epsilon}\bigg|_{\epsilon \to 0+}, \quad D_c(v_0 s_{12}) \to -\left(\frac{i}{4\pi^2}\right)\frac{1}{(s_1 - s_2 - i\bar{\epsilon})^2}\bigg|_{\bar{\epsilon} \to 0+},$$

and, as evaluated in HMF#1, Chapter 8, Section B, the corresponding, self-linkage computation for linkages by a scalar field $\pi(x)$ yields the factors $e^{is\Lambda^2 \frac{\alpha_G}{\pi}} \cdot e^{-\frac{\alpha_G}{\pi} \ln(\frac{\Lambda^2}{m^2})} \cdot (sm^2)^{-\alpha_G/\pi}$, with momentum cut-off $\Lambda^2 \sim (\bar{\epsilon})^{-1}$. In sequence, these terms correspond to a model-dependent mass renormalization, a wave-function renormalization, and a damping of the s-integrand for large s. It is this latter factor which is of interest here, which damping remains after the $s \to -i\tau$ variable change introduced above. The model is not particularly realistic; but it again displays damping at large s, or τ, values.

Let us therefore assume that, in general, such large-τ damping does result from previously-neglected self-linkages along each line; and take the simplifying step of inserting an effective, upper cut-off q/m^2 in the τ-integral of (8.32), corresponding to the largest value of τ that enters when self-linkages are included,

$$\int_0^{q/m^2} \frac{d\tau}{\tau} \tau^n \cdot e^{-m^2\tau - b^2/4\tau} \equiv 2\left(\frac{b}{2m}\right)^n H_n(mb). \tag{8.34}$$

In effect, $K_n(mb)$ will then be replaced by $H_n(mb)$, a real, positive quantity with an upper bound given by

$$H_n(mb) < \frac{1}{2}\left(\frac{2q}{mb}\right)^n \int_0^q \frac{dt}{t}\, e^{-t - (mb)^2/4t}.$$ (8.35)

What is the dimensionless quantity q? It can depend on the relevant, renormalized parameters of the theory, an effective α_G, m, and b. If q is chosen as a constant, q_0, then the summation of (8.33) will yield an approximate factor of $s^{(q_0^2 \alpha_G / 8\pi (mb)^2)} \cdot \bar{K}_0^2(mb)$, where $\bar{K}_0$ differs from K_0 in that its defining integral has an upper limit q_0, rather than ∞. But it is easy to show that, for $mb \gg 1$, there is (exponentially) little difference between $\bar{K}_0$ and K_0, so that an argument similar to that leading to (8.21) produces the quantity $\frac{q_0^2 \alpha_G}{8\pi(mb_0)^2}\ln(s/m^2) \sim 2mb_0$, so that $b_0 \sim [\ln(s/m^2)]^{1/3}$, and $\sigma_{TOT}(s) \sim [\ln(s/m^2)]^{2/3}$.

If, however, q is assumed to grow linearly with mb, $q \sim mb$, then the same argument reproduces the old Cheng–Wu result, $\sigma_{TOT}(s) \sim \ln^2(s/m^2)$.

It seems that whichever form one adopts, for any $q \sim (mb)^i$, with $0 \le i < 1$, one will find the tower-graph prediction of a slowly-rising σ_{TOT}. Hence, the inclusion of all crossed- as well as ladder-graphs, in this model version of the tower graphs which requires strong, proper-time damping attributable to the self-linkage graphs, generates a slowly-rising $\sigma_{TOT}(s)$; and for one special choice of cut-off, $q \sim mb$, it reproduces the form of the original Cheng–Wu result. This model is surely crude – and was so from the beginning – but crudeness does not necessarily preclude correctness; and the predictions of the next section could, conceivably, lead to qualitatively-correct Physics.

For $mb \ll 1$, the transverse cut-off γ should be taken as the inverse of an appropriate mass, and not as the smaller impact parameter; that is, γ should always be chosen as the largest, relevant quantity with the dimension of length. In this region, the corresponding sum of the tower-graph contributions to the eikonal of (8.33) does not appear to converge, since the leading term of $K_n(mb)$ is $\sim (1/2)(n-1)!/(mb/2)^n$ for small mb, a situation unchanged by the replacement of K_n by H_n if the self-linkage cut-off q can no longer be proportional to $(mb)^i$. What this indicates is that this eikonal is totally absorptive at small b; and just as in elementary, potential-theory calculations, it is to be replaced by a sufficiently large number, η, such that the entire contribution to

$$\Delta_2\sigma_{TOT} = (4\pi/m^2)\int_0^1 dx \cdot x\left[1 - e^{-\eta(s,x)}\right]$$ (8.36)

is just the "black disk" amount: $\Delta_2\sigma_{TOT} = 2\pi/m^2$. In this way, the complete tower-graph contributions for ladder and crossed-rung graphs again produce a slowly-rising σ_{TOT} with a maximum growth given by the old Cheng–Wu result. In the following section, we choose for simplicity $q \sim mb$, so that $K_n \to H_n \to H_{n_{max}} \to K_0$; but quite similar results will follow for any other choice of $q \sim (mb)^i, 0 \leq i < 1$.

8.4 Summing all the eikonal graphs

We now turn to the second unanswered question of the eikonal approach to high-energy scattering: what is the effect, as posed in this model, of summing over all the remaining $Q_N, N > 2$? From the complete formula, (7.19), one can anticipate possible cancellations from or enhancements to the tower-graph eikonal, but how severe will they be?

It is certainly possible to calculate each of the remaining Q_N from their definition, in (2.46), but enforcing the requirement of "connectedness" becomes tedious. It is much simpler to expand the RHS factor of (7.19) in powers of g^2, to perform the needed functional operations on the nth term of that expansion, and then – if possible – to sum the results. One has

$$e^{i\chi} = \sum_{n=0}^{\infty} \frac{1}{n!}(ig^2)^n \, e^{-i\sum_{a>b}\int \frac{\delta}{\delta\pi_a}D_c\frac{\delta}{\delta\pi_b}} \left(\int f_I\bar{\Delta}_c[\pi_a]f_{II}\right)\cdots\left(\int f_I\bar{\Delta}_c[\pi_n]f_{II}\right)\Bigg|_{\pi_l\to 0},$$
(8.37)

with the multiple factors of $(\int f_I\bar{\Delta}_c f_{II})$ occurring a total of n times; (8.37) is the form that this nth functional approximation takes when all radiative corrections along each NVM are suppressed, and are included when each $K_n \to K_0$.

For clarity, we work out the $n = 3$ term, state the form of the $n = 4$ term, and then infer the general result. Using the notation of the previous section, this is

$$\frac{1}{3!}\left(\frac{ig^2}{16\pi^2}(p_1 \cdot p_2)\right)^3 \int_0^\infty \frac{ds_a}{s_a^2} \int_0^\infty \frac{ds_b}{s_b^2} \int_0^\infty \frac{ds_c}{s_c^2} \, e^{-im^2(s_a+s_b+s_c)}$$

$$\cdot \iint_{-\infty}^{+\infty} d\bar{s}_a \, d\bar{t}_a \iint_{-\infty}^{+\infty} d\bar{s}_b \, d\bar{t}_b \iint_{-\infty}^{+\infty} d\bar{s}_c \, d\bar{t}_c$$

$$\cdot \exp\left[\frac{i}{4s_a}(z_{12} - \bar{s}_a p_1 + \bar{t}_a p_2)^2 + \frac{i}{4s_b}(z_{12} - \bar{s}_b p_1 + \bar{t}_b p_2)^2\right.$$

$$\left. + \frac{i}{4s_c}(z_{12} - \bar{s}_c p_1 + \bar{t}_c p_2)^2\right]$$

$$\cdot e^{\sum_{a>b}\mathcal{D}_{ab}} \cdot e^{-iG[s_a\int_0^1 d\lambda_a\pi_a(\xi_a^{(-)}(\lambda_a))+s_b\int_0^1 d\lambda_b\pi_b(\xi_b^{(-)}(\lambda_b))+s_c\int_0^1 d\lambda_c\pi_c(\xi_c^{(-)}(\lambda_c))]}\Big|_{\pi_l\to 0},$$
(8.38)

where $\mathcal{D}_{ab} = -i \int \frac{\delta}{\delta \pi_a} D_c \frac{\delta}{\delta \pi_b}$. As before, each linkage operator – and there are now $n(n-1)/2 = 3$ of them – generates a term of the form

$$\exp\left[-m^2 \alpha_G \left(\frac{s_a s_b}{4\pi \gamma^2} \right) \ln(s/m^2) \right]$$

– the result of (8.30), but without the factor of (-1) – while the $\int d\bar{s}_a \cdots \int d\bar{t}_c$ integrations remove all non-transverse $(z_{12})^2$ dependence, generating

$$\left(\frac{8\pi}{s} \right)^3 \cdot s_a \cdot s_b \cdot s_c \cdot e^{\frac{1}{4}b^2(\frac{1}{s_a}+\frac{1}{s_b}+\frac{1}{s_c})},$$

so that this $n = 3$ contribution to $\exp[i\chi]$ produces

$$(-i\alpha_g)^3 \int\int\int_0^\infty \frac{ds_a}{s_a} \cdot \frac{ds_b}{s_b} \cdot \frac{ds_c}{s_c} e^{-im^2(s_a+s_b+s_c)+\frac{ib^2}{4}(\frac{1}{s_a}+\frac{1}{s_b}+\frac{1}{s_c})}$$

$$\cdot \exp\left[-\frac{m^2 \alpha_G}{4\pi \gamma^2}(s_a s_b + s_b s_c + s_c s_a) \ln(s/m^2) \right].$$

Expanding each of the $n(n-1)/2$ factors of the form $\exp[-\frac{m^2 \alpha_G}{4\pi \gamma^2} \cdot s_a s_b \cdot \ln(s/m_2)]$, one obtains

$$\frac{(-2i\alpha_g)^3}{3!} \sum_{n_{1,2,3}=0}^\infty \frac{\left[\frac{\alpha_G}{8\pi} \frac{b^2}{\gamma^2} \ln(s/m^2) \right]^{n_1+n_2+n_3}}{n_1! n_2! n_3!} K_{n_1+n_3}(mb) \cdot K_{n_1+n_2}(mb) \cdot K_{n_2+n_3}(mb).$$

$$(8.39)$$

In contrast, for $n = 4$, one would find $n(n-1)/2 = 6$ summations over $n_1, \ldots, n_6$,

$$\frac{(-2i\alpha_g)^4}{4!} \sum_{n_1,\ldots,n_6=0}^\infty \frac{\left[\frac{\alpha_G}{8\pi} \cdot \frac{b^2}{\gamma^2} \cdot \ln(s/m^2) \right]^{\sum_{i=1}^6 n_i}}{n_1! \ldots n_6!} K_{N_1}(mb) K_{N_2}(mb) K_{N_3}(mb) K_{N_4}(mb),$$

$$(8.40)$$

where $N_i = \sum_{j=1}^6 n_j - n_i$.

Can one sum the series of these terms? Again, there is a great simplification for $\gamma = b$, $mb > 1$, and $K_n(mb) \Rightarrow [\pi/2mb]^{-1/2} \exp[-mb]$; and in this way, (8.39) becomes

$$\frac{(-2i\alpha_g)^3}{3!} \left(\frac{\pi}{2mb} \right)^{3/2} e^{-3mb} \cdot (s/m^2)^{3\alpha_G/8\pi},$$

while (8.40) yields

$$\frac{(-2i\alpha_g)^4}{4!} \left(\frac{\pi}{2mb} \right)^{4/2} e^{-4mb} (s/m^2)^{6\alpha_G/8\pi},$$

and the general pattern is clear:

$$e^{i\chi}|_{mb>1} \simeq 1 + i\chi_1 + \sum_{n=2}^{\infty} \frac{(-2i\alpha_g)^n}{n!} \left(\frac{\pi}{2mb}\right)^{n/2} e^{-nmb}(s/m^2)^{\frac{n(n-1)}{2} \cdot \frac{\alpha_G}{8\pi}}.$$

(8.41)

The next task is to give a meaningful value to the unusual sum of (8.41). Because of the factors $(-i)^n$, there will be an alternation in the signs of the terms that comprise Re exp$[i\chi]$; and because the s-dependence has the form $s^{an(n-1)/2}$, these alternations can be important. To see this in detail, we make use of a representation[6] expressed in terms of a convergent power series,

$$\sum_{n=2}^{\infty} \frac{(-iZ)^n}{n!} \equiv F(Z) = e^{-iZ} - 1 + iZ, \qquad Z = \frac{\bar{x}}{y} e^{2\alpha\beta},$$

and note that

$$S(\bar{x}, y) \equiv \frac{1}{\sqrt{\pi}} \int_{-\infty}^{+\infty} d\alpha\, e^{-\alpha^2} F\left(\frac{\bar{x}}{y} e^{2\alpha\beta}\right) = \sum_{n=2}^{\infty} \left(-i\frac{\bar{x}}{y}\right)^n \frac{1}{n!} e^{n^2\beta^2}.$$

This is the same series as that of (8.41), if the identifications

$$y = \left(\frac{s}{m^2}\right)^{\frac{\alpha_G}{16\pi}}, \qquad \beta^2 = \ln y, \qquad \bar{x} = \alpha_g \sqrt{\frac{2\pi}{mb}} \cdot e^{-mb}$$

are made, and we may therefore replace (8.41) by the integral

$$e^{i\chi}|_{mb>1} \rightarrow 1 + i\chi_1 + \frac{1}{\sqrt{\pi}} \int_{-\infty}^{+\infty} d\alpha\, e^{-\alpha^2} \left\{ e^{-i\left(\frac{\bar{x}}{y} e^{2\alpha\beta}\right)} - 1 + i\frac{\bar{x}}{y} e^{2\alpha\beta} \right\},$$

(8.42)

so that the contribution of (8.42) to σ_{TOT} is given by

$$\Delta_1 \sigma_{\text{TOT}} = \frac{2}{\sqrt{\pi}} \int_{1/m}^{\infty} d^2b \int_{-\infty}^{+\infty} d\alpha\, e^{-\alpha^2} \left[1 - \cos\left(\frac{\bar{x}}{y} e^{2\alpha\beta}\right) \right].$$

(8.43)

It will be convenient to make the variable change: $\alpha \rightarrow \frac{1}{2\beta} \ln(\frac{y}{\bar{x}} z)$. Approximating $\ln(\bar{x})$ by $-mb$, and with the additional variable change $mb = \beta^2(u-1)$, and with $x = 4\beta^2$, one finds

$$\Delta_1 \sigma_{\text{TOT}} = \frac{\sqrt{\pi}}{4m^2} x^{3/2} \int_{1+4/x}^{\infty} du(u-1)\, e^{-\frac{xu^2}{16}} \int_0^{\infty} dz \frac{(1-\cos(z))}{z^{1+u/2}} e^{-\frac{1}{x} \ln^2 z},$$

(8.44)

where $\Delta_1 \sigma_{\text{TOT}}$ is again that contribution to σ_{TOT} arising from $b > 1/m$. These integrals are real and positive definite, but cannot be evaluated analytically, so that a numerical approach is necessary. However, it is fairly easy to approximate (8.44), so that the evaluation can be done in an analytic manner, and for this one

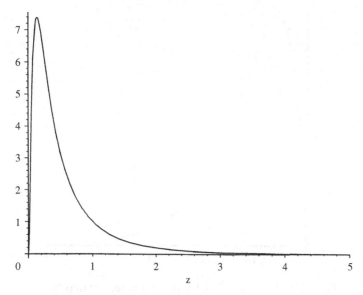

Fig. 8.1 A plot of $\exp[-x^{-1}(\ln(z))^2 - (1 + u/2)\ln(z)]$ vs z for $x = u = 2$.

first rewrites the z-integral of (8.44) as $\int_0^\infty dz \exp[-f(z)][1 - \cos(z)]$, where $f(z) = -[\frac{1}{x}(\ln z)^2 + (1 + \frac{u}{2})\ln(z)]$, and where a typical graph of $\exp[-f(z)]$, for $u = x = 2$, is displayed in Fig. 8.1. The peak of that curve occurs at a value $z_0 < 1$, and its peak height is $\exp[-f(z_0)]$. For $z > z_0$, the curve has a rough, Gaussian appearance, but for z values less than z_0 this is not the case, for the curve vanishes rapidly as $z \to 0$. However, $\exp[-f(z)]$ is multiplied by $2 \cdot \sin^2(z/2)$, which vanishes as $z \to 0$; and therefore it is not too inaccurate to replace $\exp[-f(z)]$ by a simpler Gaussian form about z_0, since the contributions for $z < z_0$ are going to be very small; and we will therefore write $f(z) \simeq f(z_0) + (1/2)(z - z_0)2f''(z_0)$, with z_0 determined by the condition $f'(z_0) = 0 : z_0 = \exp[-(x/2) \cdot (1 + u/2)]$. The argument of this exponential factor is, following from the lower limit of the u-integral, always more negative than $-(1 + x/2)$, so that z_0 is always small, expecially for large values of x. If no further approximation is used, the $\int dz$ would be given in terms of probability integrals, $\Phi(x)$.

But the smallness of z_0, especially in a region where the vanishing of $[1 - \cos(z)]$ removes most of the error, now suggests an additional approximation: replace $\int_0^\infty dz$ by $(1/2)\int_{-\infty}^{+\infty} dz$, so that the integral can be evaluated[7] immediately as

$$\frac{1}{2}\int_{-\infty}^{+\infty} dz' = \frac{1}{2}e^{-f(z_0)}\sqrt{\frac{2\pi}{f''(z_0)}} \cdot \left[1 - e^{-\frac{1}{2f''(z_0)}} \cdot \cos z_0\right], \quad (8.45)$$

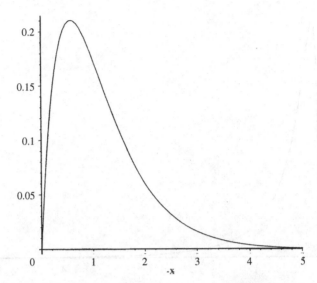

Fig. 8.2 A plot of $[(me)^2/2\pi]\sigma_{\text{TOT}}(s)$ vs $x = (\alpha_G/4\pi)\ln(s/m^2)$.

which is a convergent integrand for large u. Further, since $z_0 < 1$, expansion of the $\cos(z_0)$ of (8.45) will give corrections to the result obtained by setting $z_0 \to 0$, which are smaller by exponential dependence on x; and hence it is legitimate to replace $\cos(z_0)$ in (8.45) by unity.

Although the resulting integrand is reminiscent of that of (8.20), the technique used for the approximate evaluation of the latter will not work here; rather, because $u_{\min} = 1 + 4/x$, and $x > 0$, $h(u, x) = (x/4)\exp[-x(1 + u/2)]$ is < 1 for any value of x, and the terms of (8.45) may be expanded in powers of $h(u, x)$, with the linear term in h generating

$$\Delta_1\sigma_{\text{TOT}}(s) \simeq \left(\frac{2\pi}{m^2 e^2}\right) \cdot x\, e^{-7x/4}, \qquad x = \left(\frac{\alpha_G}{4\pi}\right)\ln(s/m^2), \qquad (8.46)$$

and with corrections which are smaller by exponential factors of x.

The graph of (8.46) in Fig. 8.2 tells the entire story. As x increases from zero, $\Delta_1\sigma_{\text{TOT}}$ rises linearly with $\ln(s)$, peaks at $x = 4/7$, and then falls off exponentially. There are therefore, as suspected, sufficient cancellations within this model to remove the rising total cross section of the tower-graphs. The details of the calculation are certainly subject to correction; but the result of (8.46) does suggest serious cancellations away from the tower-graph result. If, for example, one imagines that scattering at $s \sim 1\,\text{TeV}^2$ corresponds to values of $x \sim 0.3$, then the peak of $\Delta_1\sigma_{\text{TOT}}$ should appear at $x \sim 0.6$, corresponding to a doubling of the value of $\ln(s)$. Such an energy is probably higher than that of the so-called "cosmic ray point", and if so there is little possibility of its ever being measured directly.

Without any further calculation, one expects the $mb < 1$ contributions to reduce (as for the tower-graphs, but more quickly) to the same value of $\Delta_2 \sigma_{TOT} = 2\pi/m^2$, which result has no real bearing on the question of σ_{TOT} at asymptotic energies.

Could this model possibly apply to QCD? Not directly, for scalar pions are neither gluons nor closed quark-loops. But if one's imagination is allowed free rein, and if scalar pions are replaceable by gluons (as the basic elements of gluonic jets), then one might imagine that timelike renormalization effects of the asymptotically-free QCD could decrease α_G by a factor of $\ln(s/m^2)$, so that $x = (\alpha_G/4\pi)\ln(s/m^2)$ can never increase significantly past a constant amount, say x_{max}; and hence that σ_{TOT} would become a non-zero constant, larger than $2\pi/m^2$, at x_{max}, and would stay at that value for larger values of s. Of course, this is rampant speculation, and one does not yet know the answer for real QCD. Modulo questions of mathematical rigor, the arguments of this section suggest that, at truly asymptotic energies, and if no further fields with higher-mass quanta appear, total cross sections could become constants, either large constants if asymptotic freedom holds, or smaller constants if it does not; but they need not continue to grow in the form of the Froissart bound.

Notes

1 H. M. Fried, Y. Gabellini, and J. Avan, *Eur. Phys. J. C* **13** (2000) 699.
2 The author thanks Prof. Berndt Müller for this felicitous phrase.
3 To the best of the author's knowledge, this is another unsolved problem of several decades ago: to start from first (Lagrangian) principles of QFT, and construct these simplest of all non-Abelian scattering amplitudes.
4 See, for example, A. Krisch, *Polarized Protons and Siberian Snakes*, Proceedings of the Fourth Workshop on QCD, at the American University of Paris, June 1998; published by World Scientific.
5 The author is indebted to Prof. T. T. Wu for several informative discussions and suggestions concerning the asymptotic behavior of the K_n.
6 HMF#1, Chapter 10. This representation was used, and a similar result found, by R. Blankenbecler and H. M. Fried, *Phys. Rev. D* **8** (1973) 678, in an evaluation using a model Green's function which the author believes less compelling than the present $\bar{\Delta}_c[\pi]$.
7 I. S. Gradshteyn and I. M. Ryzhik, *Tables of Integrals, Series, and Products*, Academic Press, N.Y. (1965), #3.896/2.

9

Unitary ordered exponentials

Section 1.3 provides a definition of ordered exponentials (OEs) and some algebraic analysis based on differential and/or integral equations. More mathematically-detailed descriptions, plus a wide variety of references, may be found in the Ph.D. thesis by P. Stojkov.[1] In this chapter, to illustrate the rich structure of OEs in a relatively simple, physical setting, we treat the unitary (U) OE found in the quantum-mechanical problem of the interaction of a fermion's intrinsic spin-angular momentum with a varying magnetic field. Starting from the basic differential equation (DE) in this simplest of $SU(2)$ problems, analytic approximations are constructed to this UOE – or, equivalently, to the relevant Schrödinger equation – in both the adiabatic (Section 9.2) and stochastic (Section 9.3) limits, and are compared to the exact numerical integrations. In Section 9.4, we show how "white-noise Gaussian" functional integration (FI) over our stochastic-limit approximations reproduce a result very close to the exact one; and in Section 9.5 display the connection between the IR approximation to a relevant DE and the stochastic limit of this OE. It should be emphasized that the techniques and approximations of these sections are strictly non-perturbative.

9.1 Algebraic and differential structure

The basic DE of interest here may be written as

$$\frac{dU}{dt} = i\sigma \cdot \mathbf{B}(t)U(t), \quad U(0) = 1, \quad t \geq 0, \tag{9.1}$$

and, with $\mathbf{B}(t)$ replaced by $e\mathbf{B}(t)/2mc$, corresponds to the Schrödinger equation for the wave function of a non-relativistic fermion of spin-angular momentum $\mu = e\hbar/2mc$ interacting with a varying magnetic field, $\mathbf{B}(t)$. For simplicity, we absorb the factors $e/2mc$ into the magnitude of $\mathbf{B}$, as written in (9.1), and

149

understand the corresponding solution to the Schrödinger DE to be given by $\psi(t) = U(t) \cdot \psi(0)$, where $\psi(0)$ represents the wave function at the earlier time, $t = 0$. Here, the σ_i denote the 2×2 Pauli matrices, and the $B_i(t)$ are real input functions. The unitary solution to (9.1) is

$$U(t) = \left(e^{i \int_0^t dt' \sigma \cdot \mathbf{B}(t')} \right)_+, \tag{9.2}$$

using the notation of Section 1.3. The material presented here was published[2] in 1987, with subsequent[3,4] additions; prior work on the SC adiabatic limit can be found elsewhere.[5,6]

The SC situation may be defined by the requirement $\int_0^t dt' B(t') > 1$, $B = +\sqrt{\mathbf{B}^2}$, in contrast to the weak-coupling, or perturbative regime for which one assumes the converse, $\int_0^t dt' B(t') < 1$; in the latter case it is simple to derive a valid representation for U in terms of an expansion in multiple integrals over ascending powers of $B(t')$. For the SC case, two distinct limiting regions can be defined, one for which $|d\hat{B}/dt|$ is "small" (the adiabatic, or quasistatic limit), and the opposite ("stochastic") situation for which it is large. Clearly, if $\hat{\mathbf{B}}(t) \equiv \mathbf{B}(t)/B(t)$ did not depend on time, and were fixed in one direction, a choice of coordinate axes could be made so that only one of the σ_i need appear, and the OE would become an ordinary exponential (oe) involving that σ_i. When $\hat{\mathbf{B}}(t)$ varies with t, however, the problem becomes non-trivial, and naturally divides into these two quite different limits.

By "large" or "small", one must mean the magnitude $|d\hat{\mathbf{B}}/dt|$ with respect to the only other relevant quantity of like dimension, $B(t)$; and hence if one defines $\rho(t) \equiv |\frac{d\hat{\mathbf{B}}(t)}{dt}|/B(t)$, the SC adiabatic and stochastic limits are defined by $\rho \ll 1$ and $\rho \gg 1$, respectively. The word "stochastic" might properly be replaced by "rapidly varying input"; but it is appropriate because such behavior of ρ is expected in situations where a subsequent FI is performed with a "white-noise Gaussian" weighting, as discussed in Section 9.4.

Because it is always possible by a unitary transformation to reduce this problem containing a three-dimensional $\mathbf{B}(t)$ to an equation of the form (9.1) with a two-component $\mathbf{B}(t)$, we henceforth assume that $\mathbf{B}(t)$ lies in the (x, y) plane. Some of the analysis of subsequent approximations is changed upon returning to a three-dimensional input; but, as noted in Note 2, it is not significant. It will also be convenient to express a portion of these results in terms of the representation $U = F_0 + i\sigma \cdot \mathbf{F}(t)$, with the unitarity restriction: $F_0^2 + F^2 = 1$.

9.2 The $SU(2)$ adiabatic limit

In the extreme adiabatic limit, $\rho = 0$, corresponding to $d\hat{\mathbf{B}}/dt = 0$, all the complexity of the problem disappears, for then, as noted above, one can choose an arbitrary spatial axis to lie along the direction of **B**, and the OE becomes an ordinary exponential, $U = \cos(G) + \mathrm{i}\sigma \cdot \mathbf{B}\sin(G)$, with $G(t) = \int_0^t \mathrm{d}t' B(t')$.

Suppose now that $\hat{\mathbf{B}}(t)$ is a slowly varying unit vector, in the sense of very small ρ; then it is reasonable to choose as an initial guess for $U(t)$ the same limiting form,

$$U_0(t) = \mathrm{e}^{\mathrm{i}\sigma \cdot \mathbf{Q}_0(t)}, \tag{9.3}$$

where $\hat{\mathbf{Q}}_0(t) = \hat{\mathbf{B}}(t)$ and $Q_0(t) = |\mathbf{Q}_0(t)| = \int_0^t \mathrm{d}t' B(t')$, $B = |\mathbf{B}|$. This is not correct, but it is unitary, and its deviation from the exact V can be expressed by a unitary $V(t)$: if $U(t) = U_0(t)V(t)$, with U_0 given by (9.3), then V must satisfy the exact DE

$$\frac{\mathrm{d}V}{\mathrm{d}t} = \mathrm{i}\sigma \cdot \left(\mathbf{B} - \hat{\mathbf{Q}}_0 \frac{\mathrm{d}Q_0}{\mathrm{d}t} \right) V - \mathrm{i}Q_0 \int_0^1 \mathrm{d}\mu\, \mathrm{e}^{-\mathrm{i}\mu\sigma \cdot \mathbf{Q}_0} \sigma \cdot \frac{\mathrm{d}\hat{\mathbf{Q}}_0}{\mathrm{d}t} \mathrm{e}^{+\mathrm{i}\mu\sigma \cdot \mathbf{Q}_0} V, \tag{9.4}$$

or

$$\frac{\mathrm{d}V}{\mathrm{d}t} = \mathrm{i}\sigma \cdot \mathbf{B}_1 V,$$

with

$$\mathbf{B}_1(t) = \mathbf{B} - \mathbf{Q}_0 \frac{\mathrm{d}Q_0}{\mathrm{d}t} + \frac{1}{2} \left\{ \sin(2Q_0) \frac{\mathrm{d}\hat{\mathbf{Q}}_0}{\mathrm{d}t} - [1 - \cos(2Q_0)]\hat{\mathbf{Q}}_0 \times \frac{\mathrm{d}\hat{\mathbf{Q}}_0}{\mathrm{d}t} \right\}. \tag{9.5}$$

We write (9.5) in the form $B_1 = \mathcal{B}(Q_0, \hat{\mathbf{Q}}_0; \mathbf{B})$, and note that while the first two RHS terms of (9.5) will cancel for the specific choice of Q_0 and $\hat{\mathbf{Q}}_0$, the functional form of (9.5) will be useful later on.

Under the initial condition $V(0) = 1$, the exact solution to (9.5) is the OE

$$V(t) = \left(\exp\left[\mathrm{i} \int_0^t \mathrm{d}t' \sigma \cdot \mathbf{B}_1(t') \right] \right)_+. \tag{9.6}$$

But if, in the $\rho \ll 1$ regime, the U_0 of (9.3) is a reasonable first approximation to U, then a reasonable approximation to (9.6) should be given by

$$V_1(t) = \mathrm{e}^{\mathrm{i}\sigma \cdot \mathbf{q}_1(t)}, \tag{9.7}$$

where

$$\mathbf{q}_1(t) = \hat{\mathbf{E}}_1(t), \quad q_1(t) = |\mathbf{q}_1(t)| = \int_0^t \mathrm{d}t' |\mathbf{E}_1(t')|. \tag{9.8}$$

With this approximation, we have an "improved" estimate of $U(t)$,

$$U_1(t) \equiv U_0 V_1 = \mathrm{e}^{\mathrm{i}\sigma \cdot \mathbf{Q}_0(t)} \cdot \mathrm{e}^{\mathrm{i}\sigma \cdot \mathbf{q}_1(t)}. \tag{9.9}$$

But the combination of (9.9) is unitary, and can be rewritten in a manifestly unitary form as

$$U_1(t) = e^{i\boldsymbol{\sigma}\cdot\mathbf{Q}_1(t)}, \qquad (9.10)$$

with

$$Q_1(t) \equiv |\mathbf{Q}_1(t)| = \arccos[\cos Q_0 \cdot \cos q_1 - (\hat{\mathbf{Q}}_0 \cdot \hat{\mathbf{q}}_1)\sin Q_0 \cdot \sin q_1],$$

or

$$Q_1(t) = Q(Q_0, q_1; \hat{\mathbf{Q}}_0 \cdot \hat{\mathbf{q}}_1); \qquad (9.11)$$

and

$$\hat{\mathbf{Q}}_1(t) = [\hat{\mathbf{Q}}_0 \sin Q_0 \cdot \cos q_1 + \hat{\mathbf{q}}_1 \sin q_1 \cdot \cos Q_0$$
$$+ (\hat{\mathbf{q}}_1 \times \hat{\mathbf{Q}}_0)\sin q_1 \cdot \sin Q_0] \cdot (\sin Q_1)^{-1},$$

or

$$\hat{\mathbf{Q}}_1(t) = \hat{\mathbf{Q}}(\hat{\mathbf{Q}}_0, \hat{\mathbf{q}}_1; Q_0, q_1), \qquad (9.12)$$

where the quantities Q and $\hat{\mathbf{Q}}$ are defined by (9.11) and (9.12), respectively.

But the same process can be repeated: instead of the U_0 of (9.3) we now have the U_1 of (9.10), and can define a better approximation $U_2 = \exp[i\boldsymbol{\sigma}\cdot\mathbf{Q}_2]$, with

$$\mathbf{B}_2(t) = \mathcal{B}(Q_1, \hat{\mathbf{Q}}_1; \mathbf{B}),$$

$$\hat{\mathbf{q}}_2(t) = \hat{\mathbf{B}}_2(t), \quad q_2(t) = \int_0^t dt'\,|\mathbf{B}_2(t')|,$$

$$Q_2(t) = Q(Q_1, q_2; \hat{\mathbf{Q}}_1 \cdot \hat{\mathbf{q}}_2), \quad \hat{\mathbf{Q}}_2(t) = \hat{\mathbf{Q}}(\hat{\mathbf{Q}}_1, \hat{\mathbf{q}}_2; Q_1, q_2).$$

Clearly, the process can be repeated an infinite number of times; and if it converges, can be represented by the fixed-point equations

$$Q^* = Q(Q^*, q^*; \hat{\mathbf{Q}}^* \cdot \hat{\mathbf{Q}}^*), \qquad \hat{\mathbf{Q}}^* = \hat{\mathbf{Q}}(\hat{\mathbf{Q}}^*, \hat{\mathbf{q}}^*; Q^*, q^*),$$

$$\hat{\mathbf{q}}^* = \hat{\mathcal{B}}(Q^*, \hat{\mathbf{Q}}^*; \mathbf{B}), \quad q^* = \int_0^t dt'\,|\mathcal{B}(Q^*, \hat{\mathbf{Q}}^*; \mathbf{B})|, \qquad (9.13)$$

where $Q^*, \hat{\mathbf{Q}}^*, \hat{\mathbf{q}}^*, q^*$ and $\mathbf{B}$ are functions of t, and the functional forms $Q, \hat{\mathbf{Q}}$, and $\mathcal{B}$ are given by (9.5), (9.11), and (9.12).

For an arbitrary input $\mathbf{B}(t)$, there is probably little hope of finding or proving convergence, although for some suitably simple input this might be possible. For the simplest input vector of constant magnitude B rotating in the (x, y) plane at constant angular frequency ω, where $\rho = \omega/B$, then for small $\rho = 0.1$ one can see from Fig. 9.1 that U_1 is a better approximation to the exact (numerically

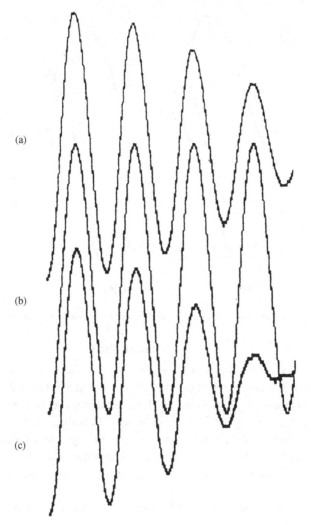

Fig. 9.1 Curves of (a) F_0, (b) $\cos Q_0$, and (c) $\cos Q_0 \cdot \cos q_1$; for the situation $\rho = 0.1$. (N.B. In this and the following figures, all curves plot the negative of every function indicated. Time increases from left to right.)

integrated) U than is U_0; there, the first two approximations to F_0, labeled (b) and (c), are to be compared with the exact result, labeled (a). Curve (b) represents $F_0 = \cos[Q_0(t)]$ for the adiabatic limit, $\rho = 0$, while (c) is the first correction to F_0, $\cos[Q_0]\cos[q_1]$. As t increases, one is "beating" two frequencies against each other, the Larmor B and the smaller ω, with the latter providing a slow modulation of the former. While (b) contains no modulation, (c) provides a bit too much, which should be corrected, in part, by the next approximation, etc.

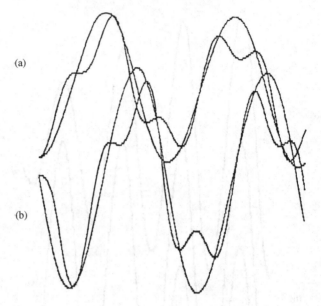

Fig. 9.2 Superpositions of (a) F_0 and $\bar{F}_0$, and (b) F_3 and $\bar{F}_3$; for $\rho = 1$.

9.3 The stochastic limit

As ρ is increased, the forms of the exact solutions change dramatically. For $\rho \sim 1$, with constant ω and B, the exact F_0 is displayed in Fig. 9.2, and bears no resemblance to its form in the adiabatic limit. As ρ is increased further, for $\rho \gg 1$, there is a great simplification, with F_0 taking the form of small, rapid, ω-oscillations superimposed upon a cosine of larger magnitude and much slower frequency $\sim B^2/2\pi\omega$. When B and ω are themselves time-dependent, the slowly-varying behavior of F_0 can become considerably more complicated than a simple cosine.

For $\rho \gg 1$, we again choose for $U(t)$ the manifestly unitary form,

$$U(t) = \exp[\mathrm{i}\boldsymbol{\sigma} \cdot \mathbf{G}(t)], \quad \mathbf{G} = \hat{\mathbf{G}}G, \quad G = +\sqrt{\mathbf{G}^2},$$

and substitute into (9.1) to obtain

$$\boldsymbol{\sigma} \cdot \mathbf{B}(t) = \int_0^1 \mathrm{d}\mu \, \mathrm{e}^{\mathrm{i}\mu\boldsymbol{\sigma}\cdot\mathbf{G}} \left(\boldsymbol{\sigma} \cdot \frac{\mathrm{d}\mathbf{G}}{\mathrm{d}t} \right) \mathrm{e}^{-\mathrm{i}\mu\boldsymbol{\sigma}\cdot\mathbf{G}}, \tag{9.14}$$

or

$$\mathbf{B}(t) = \hat{\mathbf{G}}\frac{\mathrm{d}G}{\mathrm{d}t} - \frac{1}{2}[1 - \cos(2G)]\left(\hat{\mathbf{G}} \times \frac{\mathrm{d}\hat{\mathbf{G}}}{\mathrm{d}t} \right) + \frac{1}{2}\sin(2G)\frac{\mathrm{d}\hat{\mathbf{G}}}{\mathrm{d}t}, \tag{9.15}$$

which is equivalent to the pair of exact relations

$$\frac{dG}{dt} = \hat{\mathbf{G}}(t) \cdot \mathbf{B}(t),$$ (9.16)

and

$$\frac{d\hat{\mathbf{G}}}{dt} = B[\hat{\mathbf{G}} \times \hat{\mathbf{B}} + (\hat{\mathbf{B}} - \hat{\mathbf{G}}(\hat{\mathbf{B}} \cdot \hat{\mathbf{G}})) \cot G].$$ (9.17)

With the initial conditions: $G(0) = 0$, $\hat{\mathbf{G}}(0) = 0$, the magnitude $G(t)$ is completely determined by $\hat{\mathbf{G}}$. For simplicity, we again assume that $\mathbf{B}$ lies in the (x, y) plane.

Since $\hat{\mathbf{G}}$ is a unit vector, it can be specified by two independent quantities, which we choose as ϕ and δ, and write, with $\tau = Bt$,

$$\hat{\mathbf{G}}(\tau) = \cos\phi(\tau) \cdot \hat{\mathbf{B}}\left(\int_0^\tau d\tau' \rho(\tau') - \delta(\tau)\right) + \hat{z}\sin\phi(\tau),$$ (9.18)

whose (x, y) projection is taken as a phase-changed $\hat{\mathbf{B}}$. The latter, a solution of the DE

$$\frac{d\hat{\mathbf{B}}}{d\tau} = \omega \times \hat{\mathbf{B}},$$

can be written as

$$\hat{\mathbf{B}}\left(\int_0^\tau \rho\, d\tau'\right) = \hat{i}\cos\left(\int_0^\tau \rho\, d\tau'\right) + \hat{j}\sin\left(\int_0^\tau \rho\, d\tau'\right).$$

Substitution of (9.18) into (9.17) yields the two independent equations

$$\frac{d\delta}{d\tau} = \rho - \tan\phi \cdot \cos\delta - \frac{\sin\delta}{\cos\phi} \cdot \cot G,$$ (9.19)

and

$$\frac{d\phi}{d\tau} = \sin\delta - \sin\phi \cdot \cos\delta \cdot \cot G,$$ (9.20)

which, together with the initial conditions $\delta(0) = \phi(0) = 0$, and the relation (following from (9.16)),

$$G(\tau) = \int_0^\tau d\tau' \cos\delta(\tau') \cdot \cos\phi(\tau'),$$ (9.21)

completely determines $\hat{\mathbf{G}}$.

The above relations are quite nonlinear, and it is difficult to have any intuition about their solutions in the large-ρ limit. To this end, we again, for (initial) simplicity, consider ω and B and $\rho = \omega/B$ as constants, and watch the exact solutions for $F_0 = \cos(G)$ change as ρ is increased. Just the "experimental" knowledge that, for $\rho \gg 1$, F_0 consists of rapid, small oscillations superimposed on a slowly varying cosine function is enough to suggest an argument that can be followed to extract the form of that slowly varying function. For, from (9.21), this means that as far as the "averaged" behavior is concerned, the quantity $J = \cos(\delta(\tau))\cos(\phi(\tau))$ can be treated as a constant (a statement which, clearly, must be re-examined when one discusses the rapid oscillations about the slowly varying solution). It will be useful to define the associated quantity $H = \cos(\phi(\tau))\sin(\delta(\tau))$, so that $\cos^2(\phi) = J^2 + H^2$, and the exact relations (9.19)–(9.21) can be expressed as

$$J' = -\rho H + [1 - J^2]\cot G, \tag{9.22}$$
$$H' = -\sin\phi + \rho J - HJ \cot G, \tag{9.23}$$

and

$$G = \int_0^\tau d\tau' J(\tau'). \tag{9.24}$$

For the "averaged" behavior, $J \sim$ constant $\equiv \xi(\rho)$, and (9.22) may be replaced by

$$H = \left(\frac{1}{\rho}\right)(1 - \xi^2)\cot G, \tag{9.25}$$

with $G \simeq \tau\xi$. Just as G depends on the slowly varying time dependence, so must the "averaged" H of (9.25). Substituting the latter into (9.23), with $G = \tau\xi$, yields an equation for an "averaged" $\sin(\phi)$,

$$\sin\phi = \rho\xi + \xi(1 - \xi^2)/\rho. \tag{9.26}$$

The form of (9.26) will be more complicated if ρ depends upon t, or τ, but for $\rho \gg 1$ this extra dependence need not be important. For the remainder of this derivation we shall continue to suppose that ρ is essentially constant; but we shall not hesitate to state the results for time-dependent ρ, where the final formulae continue to work in a satisfactory way. In the figures that follow, we denote the exact, or numerically integrated solutions, by F_0, F_i, and our approximations to them by $\bar{F}_0$, $\bar{F}_i$.

If the analysis leading to (9.26) is correct, $\sin\phi$ should display an "averaged" behavior, with rapid oscillations superimposed on a constant background; and this is true, experimentally, as one can see in Fig. 9.3. One should note that there has been a change of procedure used here, in the following sense. An exact (numerical) integration of (9.19)–(9.21) yields a value of G that never increases past π, while $\sin\phi$ and J are positive when the average G is increasing

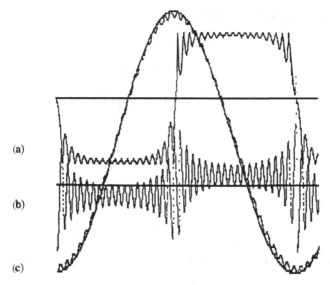

(a)

(b)

(c)

Fig. 9.3 Graphs of (a) $\sin\phi$, (b) $J = \cos\phi \cdot \cos\delta$, and (c) a superposition of F_0 and $\bar{F}_0$; for $\rho = 6$, $E = 10$.

and negative when it decreases (so that $\hat{\mathbf{G}}$ can cover all points on the unit sphere). In contrast, our "averaged" G will increase without limit, so that $\sin(G)$ may become negative (just when the exact G was decreasing), while the averaged $\sin\phi$ and J are replaced by positive constants. In this way we are able to represent the correct signs of all of the F_0, F_i. This same feature of always positive $\sin\phi$ and J can occur in numerically-integrated solutions of the exact (9.19)–(9.21), depending on the accuracy of the computation and the passage through the singular regions of $\cot(G)$. For our purposes, both $\sin\phi$ and J can be thought of as having an "averaged" constant value, even though in reality they oscillate about that value, and oscillate wildly near the regions $G \sim n\pi$. In contrast, a plot of $\sin\delta$ displays an almost uniform density of points spread over the same intervals.

We now use the "averaged" constancy of $\sin\phi$, or of $\cos^2\phi = J^2 + H^2$, to determine the dependence of ξ on ρ. For, if the averaged value of $\frac{d}{d\tau}(\cos^2\phi)$ is to vanish, from (9.22) and (9.23) one finds another expression for the averaged H,

$$0 \simeq H\sin\phi + J[1 - (J^2 + H^2)]\cot G,$$

or

$$H = \xi \sin\phi \cot(\xi\tau). \tag{9.27}$$

Comparing with (9.25) we obtain

$$\xi \sin \phi = \left(1 - \xi^2\right)/\rho, \tag{9.28}$$

and, finally, comparing (9.28) with (9.26) yields

$$(1 - \xi^2)/\rho = \rho \xi^2 [1 + (1 - \xi^2)/\rho^2],$$

or

$$\xi(\rho) = \sqrt{1 + \rho^2/4} - \rho/2. \tag{9.29}$$

In obtaining (9.29) it has been supposed that $\xi > 0$ and $1 - \xi^2 > 0$. Limiting forms are

$$\xi(\rho)|_{\rho \gg 1} \sim 1/\rho - \frac{1}{\rho^2} + \cdots, \quad \text{and} \quad \xi(\rho)|_{\rho \ll 1} \sim 1 - \rho/2 + \cdots.$$

With these relations our "averaged" solutions for F_0, F_3 are given by

$$\bar{F}_0 = \cos G, \tag{9.30}$$
$$\bar{F}_3 = \sin G, \tag{9.31}$$

where

$$G = \tau \xi(\rho) \to \int_0^\tau d\tau' \xi(\rho(\tau')) \tag{9.32}$$

is appropriate as a first generalization to time-dependent B and ω. The accuracy of these expressions is quite good for $\rho > 5$, where deviations from the numerically-integrated F_0, F_3 are rarely worse than a few percent, and frequently much less. Even for $\rho \sim 1$, where this analysis is certainly not valid, one finds that these expressions for $\bar{F}_0$ and $\bar{F}_3$ do tend to average out the then non-rapid fluctuations of the machine-integrated F_0, F_3. Some typical outputs may be seen in Figs. 9.4–9.6, including several examples of time-dependent B and ω. One finds, generally, that even if ρ has some oscillations superimposed on a constant value $\gg 1$, the $\bar{F}_0$, $\bar{F}_3$ given by (9.30)–(9.32) continue to be reasonably accurate.

Analogous approximate expressions are easily written for $F_{1,2}$. Exactly, one has

$$F_1 = \sin G \left[J \cos L + H \sin L \right], \tag{9.33}$$

and

$$F_2 = \sin G \left[J \sin L - H \cos L \right], \tag{9.34}$$

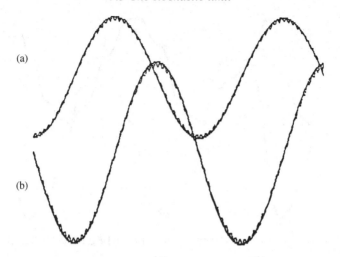

Fig. 9.4 Superpositions of (a) F_0 and $\bar{F}_0$, and (b) F_3 and $\bar{F}_3$; for $\rho = 6$, $E = 10$.

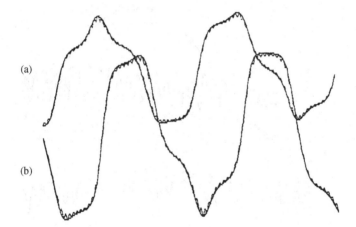

Fig. 9.5 Superpositions of (a) F_0 and $\bar{F}_0$, and (b) F_3 and $\bar{F}_3$; for $\omega = 60$, $E(t) = 10 + 5\sin(5t)$.

with $L = \int_0^t dt'\omega(t')$. Inserting the same "averaged" approximations for J, H, G as before, one finds

$$\bar{F}_1 = \xi \sin{(G + L)}, \qquad (9.35)$$

and

$$\bar{F}_2 = -\xi \cos{(G + L)}, \qquad (9.36)$$

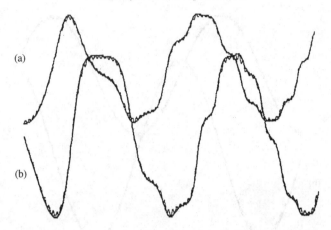

Fig. 9.6 Superpositions of (a) F_0 and $\bar{F}_0$, and (b) F_3 and $\bar{F}_3$; for $\omega = 60$, $E(t) = 10 + \cos(t^2)$.

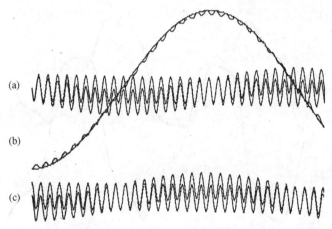

Fig. 9.7 Superpositions of (a) F_1 and $\bar{F}_1$, (b) F_0 and $\bar{F}_0$, and (c) F_2 and $\bar{F}_2$; for $\rho = 6$, $E = 10$.

where G is again given by (9.32). For large ρ, $\xi \sim 1/\rho$, and these $\bar{F}_{1,2} \sim O(1/\rho)$, in contrast to $\bar{F}_{0,3} \sim O(1)$. These $\bar{F}_{1,2}$ are therefore small, and oscillate rapidly, and should have little physical importance in any specific problem. However, as seen in Fig. 9.7, they do miss some of the slowly varying dependence of the exact $F_{1,2}$, even if the dependence is of the order of $1/\rho$.

This trouble arises in our neglect of small, rapid oscillations of J and H, because those neglected, fast oscillations could themselves be combined with similar oscillations appearing in the definition of $F_{1,2}$. We refer the interested

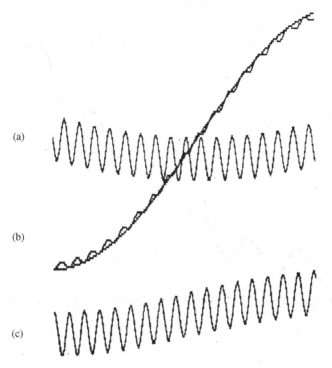

(a)

(b)

(c)

Fig. 9.8 Superpositions of (a) F_1 and $\bar{F}'_1$, (b) F_0 and $\bar{F}'_0$, and (c) F_2 and $\bar{F}'_2$; for $\omega = 60$, $E = 10$.

reader to the correction of this oversight given in Note 2, and simply write down the resulting, modified expressions for the new $\bar{F}_{1,2}$, now correct to $O(1/\rho)$:

$$\bar{F}'_1 = \xi[\sin(G + L) + \sin G], \qquad (9.37a)$$

$$\bar{F}'_2 = -\xi[\cos(G + L) - \cos G]. \qquad (9.37b)$$

The agreement between (9.37) and the exact $F_{1,2}$ is so good that on the scale used in Fig. 9.8 there is no visible difference between them. Only when the scale is enlarged to show the effects of order $1/\rho^2$ can one see the superposition of two distinct curves.

These new values of (9.37) can now be used, together with a simple unitarity argument, to produce new values of $\bar{F}'_{0,3}$ which are correct to $O(1/\rho^2)$; and, again, we refer the interested reader to Note 2 for details. The results are plotted in Figs. 9.9 and 9.10, and give a visual description of the accuracy of this effective expansion in $1/\rho$.

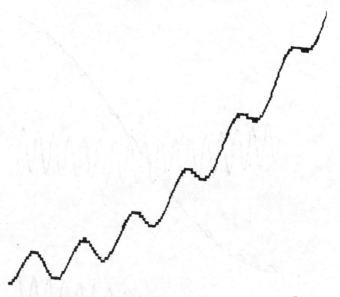

Fig. 9.9 Detail of the first shoulder of the superposition of F_0 and $\bar{F}_0'$; for $\omega = 60$, $E = 10$.

(a)

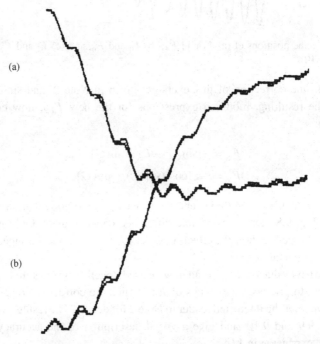

(b)

Fig. 9.10 Detail of the first shoulder for the superpositions of (a) F_0 and $\bar{F}_0'$, and (b) F_3 and $\bar{F}_3'$; for $\omega = 60$, $E(t) = 10 + 5\sin(5t)$.

9.4 Functional integration over the stochastic limit

One very nice check on the above approximations is their ability to reproduce the result of the one, non-trivial FI over an OE which can be performed analytically, that of "white-noise Gaussian" (WNG) integration over the $U(t|B)$ of (9.2). Indeed, one type of application of these results should be to stochastic FI over weightings more complicated than Gaussian. In this section, we first show why the stochastic limit is appropriate to WNG integration, and then just how closely our approximate forms reproduce the known, exact result

$$N \int d[B] \exp \left[-\frac{1}{2c} \int_0^t dt' B^2(t') \right] U(t|B) = e^{-tc}, \qquad (9.38)$$

where N is a normalization constant defined by

$$N^{-1} = \int d[B] \exp \left[-\frac{1}{2c} \int_0^t dt' B^2(t') \right].$$

In (9.38) we denote by c a real, positive constant, and continue to suppose that **B** lies in the (x, y) plane.

We first remind the reader of the derivation of (9.38). As in Chapter 2, imagine the interval $(0, t)$ broken up into n subintervals each of width $\Delta t = t/n$ and labeled by an index i, so that the **B**(t') field in each subinterval is denoted by B_i. Then, this FI may be written as

$$\text{Lim}_{n \to \infty} \prod_{i=1}^{n} N_i \int d^2 B_i \, e^{-\Delta t B_i^2/2c} \left(e^{i\Delta t \sigma \cdot \mathbf{B}} \right)_+, \qquad (9.39)$$

and the ordering of the brackets is such that terms with the larger value of i stand to the left. But each integral yields a result independent of i – that is, independent of σ – by the following argument.

Because of the Gaussian weighting, each B_i scales as $(\Delta t)^{-1/2}$; that is, in (9.39) replace each $\mathbf{B}_i$ by $\mathbf{F}_i/(\Delta t)^{1/2}$ (including the normalization, $N_i \to N_i'/(\Delta t)^{1/2}$), and for small Δt expand each $\exp[i(\Delta t)^{1/2}\sigma \cdot \mathbf{F}_i]$ so that (9.39) becomes

$$\text{Lim}_{n \to \infty} \prod_{i=1}^{n} N_i' \int d^2 F_i \, e^{-F_i^2/2c} \left(1 + i\sqrt{\Delta t}\sigma \cdot \mathbf{F}_i - \frac{\Delta t}{2} F_i^2 + \cdots \right), \qquad (9.40)$$

of which we retain only the leading, non-zero dependence proportional to Δt (the coefficient of $(\Delta t)^{1/2}$ vanishes by symmetry). Each ith integral is the same, and is trivial, yielding

$$\text{Lim}_{n \to \infty} (1 - c\Delta t)^n = e^{-ct}. \qquad (9.41)$$

The essential part of this computation has been the observation that, for WNG integration, each B_i scales as $(\Delta t)^{-1/2}$. We now consider the same FI over our "averaged" forms. The first point to be settled is whether the stochastic limit is valid, and for this we must estimate the size of $\rho^2 = (\frac{d\hat{B}}{dt})^2/B^2$. But, upon breaking up the interval $(0, t)$ into subintervals, any $\rho^2(t)$ would be replaced by

$$\rho_i^2 = (\hat{\mathbf{B}}_i - \hat{\mathbf{B}}_{i+1})^2/B_i^2(\Delta t)^2.$$

The $\hat{\mathbf{B}}_i$ dependence is of $O(1)$; but because B_i scales as $(\Delta t)^{-1/2}$, $\rho_i^2 \sim O(1/\Delta t)$, and is large. Hence the stochastic limit most certainly is relevant, and we consider the FIs of our "averaged" forms in the limit of very large ρ, $U \to \bar{F}_0 + i\sigma_3\bar{F}_3$, $\bar{F}_0 = \cos(G)$, $\bar{F}_3 = \sin(G)$. One then requires

$$N \int d[B] \exp\left[-\frac{1}{2c} \int_0^t dt' B^2(t') \right] \cdot e^{\pm iG}, \qquad (9.42)$$

which, upon writing $G \simeq \int_0^t dt' B/\rho$, and breaking up the integration region into subintervals, generates

$$\text{Lim}_{n \to \infty} \prod_{i=1}^n N_i \int d^2 B_i \, e^{-\Delta t B_i^2/2c} \cdot e^{\pm i\Delta t B_i/\rho_i}, \qquad (9.43)$$

where $\rho_i = +[(\hat{\mathbf{B}}_i - \hat{\mathbf{B}}_{i+1})^2]^{1/2}/B_i\Delta t$. Again rescaling $\mathbf{B}_i$, we now find in each subinterval both an integral over the magnitude F_i and a non-trivial angular dependence. Integration over each magnitude is immediate, leaving

$$\frac{1}{2\pi} \int_0^{2\pi} d\theta_i \left(1 \mp \frac{ic\Delta t}{|\sin(\theta_i/2)|} \right)^{-1}. \qquad (9.44)$$

The integral of (9.44) can be done exactly; with $q = \pm c \cdot \Delta t$, it is

$$1 + \frac{2iq}{\pi(1+q^2)^{1/2}} \ln\left[\left(\frac{1 - (1+q^2)^{1/2} - iq}{1 + (1+q^2)^{1/2} - iq} \right) \cdot \left(\frac{1 + (1+q^2)^{1/2}}{1 - (1+q^2)^{1/2}} \right) \right].$$

As $\Delta t \to 0$, the argument of the log becomes $\pm 2i/c\Delta t$, generating for the complete FI

$$\text{Lim}_{n \to \infty} \prod_{i=1}^n \left(1 - c\Delta t \pm \frac{2ic\Delta t}{\pi} \ln\left(\frac{2}{c\Delta t} \right) \right), \qquad (9.45)$$

which can be written as

$$e^{-ct} \cdot e^{\pm(2ict/\pi)\cdot\ln(2/c\Delta t)}\Big|_{\Delta t \to 0}. \qquad (9.46)$$

Comparison with (9.38) shows that a spurious phase has appeared, but one that can be understood, and removed, by the following argument. In every subinterval's integration, our "averaged" forms have made a small error, which is (fortunately) imaginary, and which must be removed "by hand". Instead of calculating (9.42) as we have done, we must add the proviso that we keep only the real part of every subinterval's contribution; and in this way, by not retaining and compounding the small error generated by our "averaged" forms, we can reproduce (9.38). We expect this tendency towards a spurious phase factor to appear in more complicated FIs, or in FIs that are Gaussian but not precisely in the white-noise limit; and it will be necessary to remove such spurious dependence. This can be most simply done by replacing the FI over $\exp[\pm iG]$, which we call $\langle \exp[\pm iG] \rangle$, by the quantity $[|\langle \exp(\pm iG) \rangle|^2]^{1/2}$, a computation that we henceforth label "renormalized".

More general, non-WNG weightings may be treated by calculating Gaussian fluctuations with a correlation function given by

$$\Delta_{ij}(t_1 - t_2) = \langle B_i(t_1)B_j(t_2) \rangle = \delta_{ij}(E_m/2\tau)\exp(-|t_1 - t_2|/\tau),$$

where τ is a correlation time, and E_m an appropriate magnitude. The limit $\tau^{-1} \to \infty$ for $E_m = 1$ is the WNG case, $\Delta_{ij} \to \delta_{ij}\delta(t_1 - t_2)$, while the opposite limit, $\tau^{-1} \to 0$ is effectively the adiabatic limit. (This last statement would be strictly true if ρ were defined as $|dB/dt|/B^2$, rather than as $|d\hat{B}/dt|/B$; in practice, there seems to be little difference.)

We illustrate in Figs. 9.11 and 9.12 FIs in the WNG region ($\tau^{-1} = 100$) over a variety of different possible approximations, and note that the best agreement with the exact FI is obtained by first performing the large-ρ approximation of ξ, $\xi(\rho) \sim 1/\rho$, and then performing the FI. Why this is true, and not the converse, is explained for the interested reader in Note 2; here, it is simply noted that the very close agreement of these figures, depicting WNG FI over the exact and renormalized stochastic forms is surely no accident, and is indicative of the robust nature of the stochastic approximation's $1/\rho$ expansion.

Finally, one may add that a completely independent method of finding the lowest-order terms in the $1/\rho$ expansion has been outlined in Note 4 for $SU(2)$, although it is not unambiguous for $SU(3)$. The connection here is between the first-order DE analysis discussed above, and that of the second-order DE which may be constructed to study the F_0 and F_i. In particular, when certain input functions of the second order DE are large and slowly varying, the solutions obtained are precisely those of the leading terms of the $1/\rho$ expansion of the stochastic limit. Efforts to extend these strong-coupling results to $SU(N)$, $N > 2$, have been made, but have only met with partial success, as in Note 3.

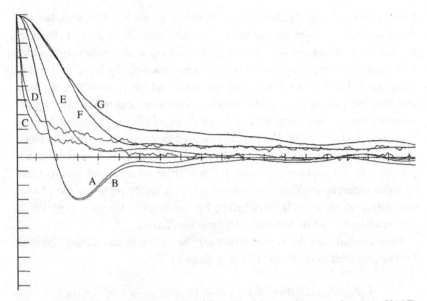

Fig. 9.11 A comparison of the result of functional integration over the exact $U(t|E)$ with several approximations, for $\tau^{-1} = 100$, $E_m = 10$, $\Delta t = 0.005$. The labeling of the curves is A = exact, numerical; B = renormalized $(1/\rho)$; $C = (1/\rho)$; D = full $\xi(\rho)$; E = renormalized, full $\xi(\rho)$; F = renormalized adiabatic; G = adiabatic.

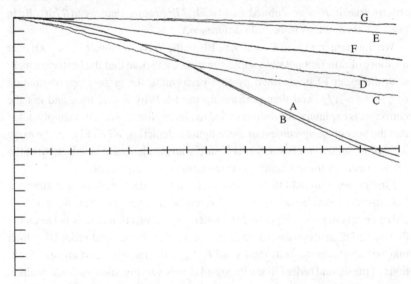

Fig. 9.12 The same comparisons, with the same labeling as in Fig. 9.11, using $\tau^{-1} = 100$, $E_m = 1$, $\Delta t = 0.005$.

Notes

1 P. Stojkov, Ph.D. Thesis: *Lasers, Heat Kernels*, and Ordered Exponentials, Physics Department, Brown University, 1998.
2 M. E. Brachet and H. M. Fried, *J. Math. Phys.* **28** (1987) 15. A brief report of parts of this work by the same authors appeared in *Physics Letters* **103A** (1984) 309.
3 H. M. Fried, *J. Math. Phys.* **28** (1987) 1275.
4 H. M. Fried, *J. Math. Phys.* **30** (1999) 1161.
5 A. Dykhne, *Sov. Phys. JETP* **11** (1960) 411; *JETP Lett.* **14** (1962) 941.
6 L. Aspinall and J. C. Percival, *Proc. R. Soc.* **90** (1967) 315, and references therein.

Index

Printed in the United States
By Bookmasters